An Introduction to Topography

An Introduction to Topography

Edited by
Royce Sutton

Larsen & Keller
www.larsen-keller.com

An Introduction to Topography
Edited by Royce Sutton
ISBN: 978-1-63549-277-4 (Hardback)

Larsen & Keller

Published by Larsen and Keller Education,
5 Penn Plaza,
19th Floor,
New York, NY 10001, USA

Cataloging-in-Publication Data

An Introduction to topography / edited by Royce Sutton.
p. cm.
Includes bibliographical references and index.
ISBN 978-1-63549-277-4
1. Physical geography. 2. Geomorphology. 3. Planetary science. 4. Geology.
5. Topographical surveying. I. Sutton, Royce.
GB60 .I58 2017
910.02--dc23

Printed and bound in China.

For more information regarding Larsen and Keller Education and its products, please visit the publisher's website www.larsen-keller.com

Table of Contents

Preface

This book elucidates the concepts and innovative models around prospective developments with respect to topography. It describes in detail the concepts and techniques related to this field. Topography refers to the study of the Earth's surface and of other celestial bodies like moons, asteroids and other planets, etc. It is generally used to make maps and is a vital part of geoscience. It also includes the study of terrains and specific landforms. This text is a compilation of chapters that discuss the most important concepts in the field of topography. It unfolds some innovative aspects which will be crucial for the holistic understanding of the subject matter. While understanding the long-term perspectives of the topics, the book makes an effort in highlighting their impact as a modern tool for the growth of the discipline. This textbook is an essential guide for both students.

Given below is the chapter wise description of the book:

Chapter 1- The study of the shape and the features of the Earth is known as topography. Research in topography is made for a number of reasons; some of the reasons are military planning and geological exploration. It is closely related to geoscience and cartography. The following text helps the reader in developing an in- depth understanding of topography.

Chapter 2- Ocean surfaces have highs and lows; the depiction of these is done with the help of ocean surface topography. Topography of the moon, terrain, elevation, pit-and-mound topography and phytogeomorphology are the additional major aspects of topography. The following content unfolds its crucial aspects in a critical and yet systematic manner.

Chapter 3- Landforms are features of the solid surface of the Earth. The arrangement of landforms is referred to as topography. Typically, landforms include hills, mountains, canyons, valleys and mid-ocean ridges. The topics discussed in the chapter are of great importance to broaden the existing knowledge on landforms.

Chapter 4- Surveying is a method that helps in the determination of terrestrial positioning of the points and the distances between them. These points are usually used to establish land maps and boundaries. Like surveying the other techniques of topography are remote sensing, satellite imagery, topographic prominence, aerial photography, bathymetry, geovisualization etc. This text examines the methods of topography in a critical manner providing key analysis to the subject matter.

Chapter 5- Cartography is the study of creating and applying maps. The types of maps mentioned in this section are raised-relief maps, pictorial maps, fantasy maps, city maps and geologic maps. The other aspects of cartography examined are cartographic labeling, cartographic generalization, map projection and contour line. This section is an overview of the subject matter incorporating all the major aspects of cartography.

Chapter 6- Topography is an interdisciplinary subject. To develop a better understanding of topography, its allied fields also need to be contemplated. Earth science studies the planet Earth which includes the interior as well as the exterior of the planet. Planetary science is the study of planets and the process that helps in their formation. This chapter will provide a glimpse of the allied fields of topography.

At the end, I would like to thank all those who dedicated their time and efforts for the successful completion of this book. I also wish to convey my gratitude towards my friends and family who supported me at every step.

Editor

1

Introduction to Topography

The study of the shape and the features of the Earth is known as topography. Research in topography is made for a number of reasons; some of the reasons are military planning and geological exploration. It is closely related to geoscience and cartography. The following text helps the reader in developing an in- depth understanding of topography.

Topography is the study of the shape and features of the surface of the Earth and other observable astronomical objects including planets, moons, and asteroids. The topography of an area could refer to the surface shapes and features themselves, or a description (especially their depiction in maps).

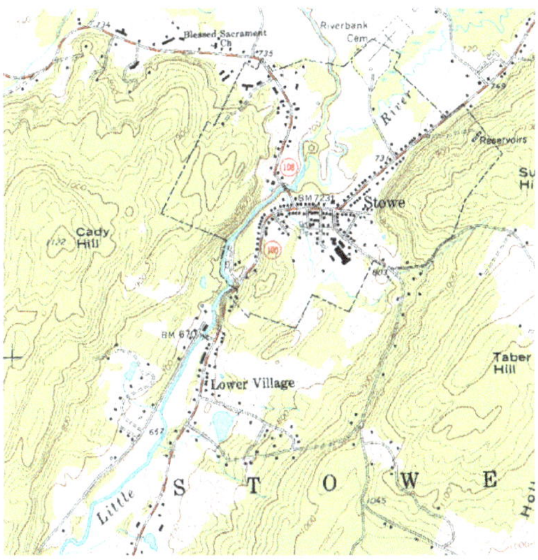

A topographic map with contour intervals

This field of geoscience and planetary science is concerned with local detail in general, including not only relief but also natural and artificial features, and even local history and culture. This meaning is less common in the United States, where topographic maps with elevation contours have made "topography" synonymous with relief. The older sense of topography as the study of place still has currency in Europe.

Topography in a narrow sense involves the recording of relief or terrain, the three-dimensional quality of the surface, and the identification of specific landforms. This is also known as geomorphometry. In modern usage, this involves generation of elevation data in digital form (DEM). It is often considered to include the graphic representation of the landform on a map by a variety of techniques, including contour lines, hypsometric tints, and relief shading.

Satellite imagery illustrating topography of the urban core of the New York City Metropolitan Area, with Manhattan Island at its center.

Etymology

The term *topography* originated in ancient Greece and continued in ancient Rome, as the detailed description of a place. The word comes from the Greek τόπος (*topos*, "place") and -γραφία (*-graphia*, "writing"). In classical literature this refers to writing about a place or places, what is now largely called 'local history'. In Britain and in Europe in general, the word topography is still sometimes used in its original sense.

Detailed military surveys in Britain (beginning in the late eighteenth century) were called Ordnance Surveys, and this term was used into the 20th century as generic for topographic surveys and maps. The earliest scientific surveys in France were called the Cassini maps after the family who produced them over four generations. The term "topographic surveys" appears to be American in origin. The earliest detailed surveys in the United States were made by the "Topographical Bureau of the Army," formed during the War of 1812, which became the Corps of Topographical Engineers in 1838. After the work of national mapping was assumed by the U.S. Geological Survey in 1878, the term topographical remained as a general term for detailed surveys and mapping programs, and has been adopted by most other nations as standard.

In the 20th century, the term topography started to be used to describe surface description in other fields where mapping in a broader sense is used, particularly in medical fields such as neurology.

Objectives

An objective of topography is to determine the position of any feature or more generally any point in terms of both a horizontal coordinate system such as latitude, longitude, and altitude. Identifying (naming) features, and recognizing typical landform patterns are also part of the field.

A topographic study may be made for a variety of reasons: military planning and geological exploration have been primary motivators to start survey programs, but detailed information about terrain and surface features is essential for the planning and construction of any major civil engineering, public works, or reclamation projects.

Techniques of Topography

There are a variety of approaches to studying topography. Which method(s) to use depend on the scale and size of the area under study, its accessibility, and the quality of existing surveys.

Direct Survey

Surveying helps determine accurately the terrestrial or three-dimensional space position of points and the distances and angles between them using leveling instruments such as theodolites, dumpy levels and clinometers.

A surveying point in Germany

Work on one of the first topographic maps was begun in France by Giovanni Domenico Cassini, the great Italian astronomer.

Even though remote sensing has greatly sped up the process of gathering information, and has allowed greater accuracy control over long distances, the direct survey still provides the basic control points and framework for all topographic work, whether manual or GIS-based.

In areas where there has been an extensive direct survey and mapping program (most of Europe and the Continental US, for example), the compiled data forms the basis of basic digital elevation datasets such as USGS DEM data. This data must often be "cleaned" to eliminate discrepancies between surveys, but it still forms a valuable set of information for large-scale analysis.

The original American topographic surveys (or the British "Ordnance" surveys) involved not only recording of relief, but identification of landmark features and vegetative land cover.

Remote Sensing

Remote sensing is a general term for geodata collection at a distance from the subject area.

Passive Sensor Methodologies

Besides their role in photogrammetry, aerial and satellite imagery can be used to identify and delineate terrain features and more general land-cover features. Certainly they have become more and more a part of geovisualization, whether maps or GIS systems. False-color and non-visible spectra imaging can also help determine the lie of the land by delineating vegetation and other land-use information more clearly. Images can be in visible colours and in other spectrum

Photogrammetry

Photogrammetry is a measurement technique for which the co-ordinates of the points in 3D of an object are determined by the measurements made in two photographic images (or more) taken starting from different positions, usually from different passes of an aerial photography flight. In this technique, the common points are identified on each image. A line of sight (or ray) can be built from the camera location to the point on the object. It is the intersection of its rays (triangulation) which determines the relative three-dimensional position of the point. Known control points can be used to give these relative positions absolute values. More sophisticated algorithms can exploit other information on the scene known a priori (for example, symmetries in certain cases allowing the rebuilding of three-dimensional co-ordinates starting from one only position of the camera).

Active Sensor Methodologies

Satellite RADAR mapping is one of the major techniques of generating Digital Elevation Models. Similar techniques are applied in bathymetric surveys using sonar to determine the terrain of the ocean floor. In recent years, LIDAR (Light Detection And Ranging), a remote sensing technique that uses a laser instead of radio waves, has increasingly been employed for complex mapping needs such as charting canopies and monitoring glaciers.

Forms of Topographic Data

Terrain is commonly modelled either using vector (triangulated irregular network or TIN) or gridded (Raster image) mathematical models. In the most applications in environmental sciences, land surface is represented and modelled using gridded models. In civil engineering and entertainment businesses, the most representations of land surface employ some variant of TIN models. In geostatistics, land surface is commonly modelled as a combination of the two signals – the smooth (spatially correlated) and the rough (noise) signal.

In practice, surveyors first sample heights in an area, then use these to produce a Digital Land Surface Model in the form of a TIN. The DLSM can then be used to visualize terrain, drape remote sensing images, quantify ecological properties of a surface or extract land surface objects. Note that the contour data or any other sampled elevation datasets are not a DLSM. A DLSM implies that elevation is available continuously at each location in the study area, i.e. that the map represents a complete surface. Digital Land Surface Models should not be confused with Digital Surface Models, which can be surfaces of the canopy, buildings and similar objects. For example, in the case of surface models produces using the lidar technology, one can have several surfaces – starting from the top of the canopy to the actual solid earth. The difference between the two surface models can then be used to derive volumetric measures (height of trees etc.).

Raw Survey Data

Topographic survey information is historically based upon the notes of surveyors. They may derive naming and cultural information from other local sources (for example, boundary delineation may be derived from local cadastral mapping). While of historical interest, these field notes inherently include errors and contradictions that later stages in map production resolve.

Remote Sensing Data

As with field notes, remote sensing data (aerial and satellite photography, for example), is raw and uninterpreted. It may contain holes (due to cloud cover for example) or inconsistencies (due to the timing of specific image captures). Most modern topographic mapping includes a large component of remotely sensed data in its compilation process.

Topographic Mapping

A map of Europe using **elevation modeling**

In its contemporary definition, topographic mapping shows relief. In the United States, USGS topographic maps show relief using contour lines. The USGS calls maps based on topographic surveys, but without contours, "planimetric maps."

These maps show not only the contours, but also any significant streams or other bodies of water, forest cover, built-up areas or individual buildings (depending on scale), and other features and points of interest.

While not officially "topographic" maps, the national surveys of other nations share many of the same features, and so they are often called "topographic maps."

Existing topographic survey maps, because of their comprehensive and encyclopedic coverage, form the basis for much derived topographic work. Digital Elevation Models, for example, have often been created not from new remote sensing data but from existing paper topographic maps. Many government and private publishers use the artwork (especially the contour lines) from existing topographic map sheets as the basis for their own specialized or updated topographic maps.

Topographic mapping should not be confused with geologic mapping. The latter is concerned with underlying structures and processes to the surface, rather than with identifiable surface features.

Digital Elevation Modeling

Relief map: Sierra Nevada Mountains, Spain

The digital elevation model (DEM) is a raster-based digital dataset of the topography (hypsometry and/or bathymetry) of all or part of the Earth (or a telluric planet). The pixels of the dataset are each assigned an elevation value, and a header portion of the dataset defines the area of coverage, the units each pixel covers, and the units of elevation (and the zero-point). DEMs may be derived from existing paper maps and survey data, or they may be generated from new satellite or other remotely sensed radar or sonar data.

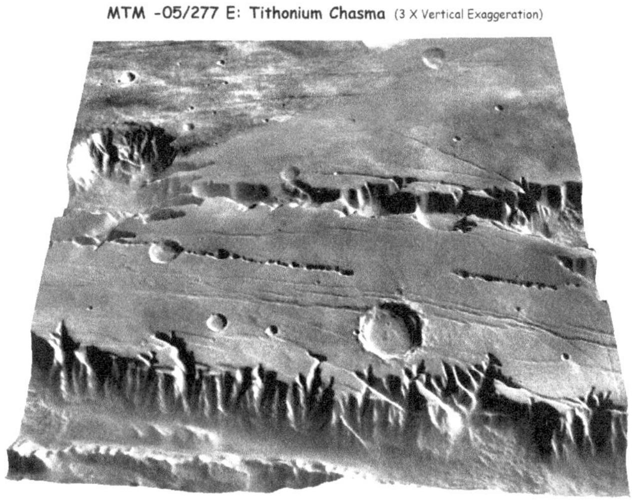

3D rendering of a DEM used for the topography of Mars

Topological Modeling

A geographic information system (GIS) can recognize and analyze the spatial relationships that exist within digitally stored spatial data. These topological relationships allow complex spatial modelling and analysis to be performed. Topological relationships between geometric entities traditionally include adjacency (what adjoins what), containment (what encloses what), and proximity (how close something is to something else).

- reconstitute a sight in synthesized images of the ground,
- determine a trajectory of overflight of the ground,
- calculate surfaces or volumes,
- trace topographic profiles,

Topography in Other Fields

Topography has been applied to different science fields. In neuroscience, the neuroimaging discipline uses techniques such as EEG topography for brain mapping. In ophthalmology, corneal topography is used as a technique for mapping the surface curvature of the cornea. In tissue engineering, atomic force microscopy is used to map nanotopography.

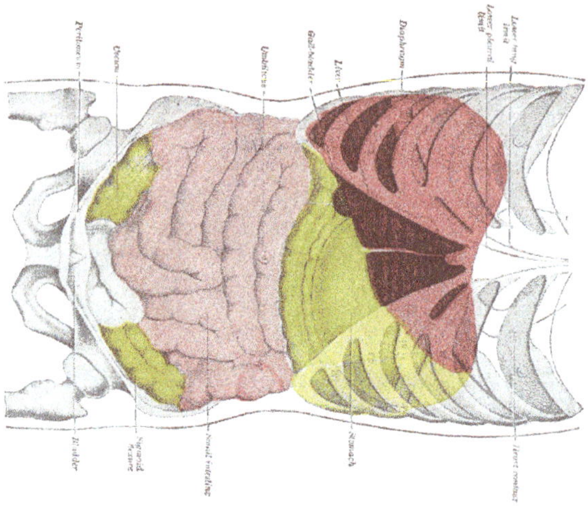

Topography of thoracic and abdominal viscera.

In human anatomy, topography is superficial human anatomy.

In mathematics the concept of topography is used to indicate the patterns or general organization of features on a map or as a term referring to the pattern in which variables (or their values) are distributed in a space.

2

Major Aspects of Topography

Ocean surfaces have highs and lows; the depiction of these is done with the help of ocean surface topography. Topography of the moon, terrain, elevation, pit-and-mound topography and phytogeomorphology are the additional major aspects of topography. The following content unfolds its crucial aspects in a critical and yet systematic manner.

Topography of the Moon

The topography of the Moon has been measured by the methods of laser altimetry and stereo image analysis, most recently from data obtained during the Clementine mission. The most visible topographic feature is the giant far side South Pole-Aitken basin, which possesses the lowest elevations of the Moon. The highest elevations are found just to the north-east of this basin, and it has been suggested that this area might represent thick ejecta deposits that were emplaced during an oblique South Pole-Aitken basin impact event. Other large impact basins, such as the maria Imbrium, Serenitatis, Crisium, Smythii, and Orientale, also possess regionally low elevations and elevated rims.

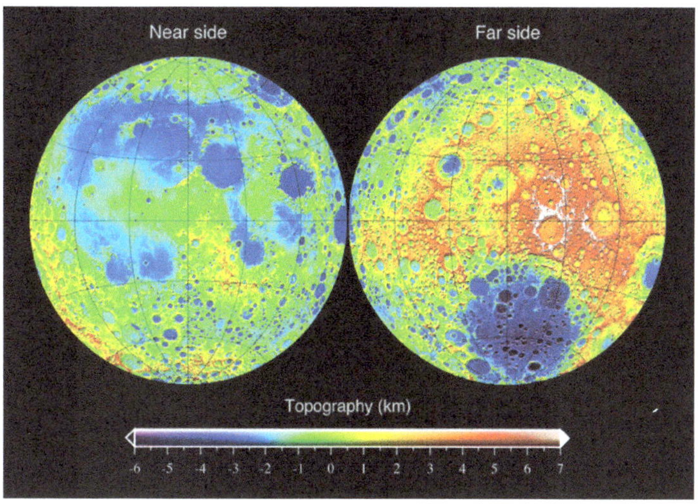

Topography of the Moon.

Another distinguishing feature of the Moon's shape is that the elevations are on average about 1.9 km higher on the far side than the near side. If it is assumed that the crust is in isostatic equilibrium, and that the density of the crust is everywhere the same, then the higher elevations would be associated with a thicker crust. Using gravity, topography and seismic data, the crust is thought to be on average about 50 ± 15 km thick, with the far-side crust being on average thicker than the near side by about 15 km.

The Moon

Lunar near side

Lunar far side

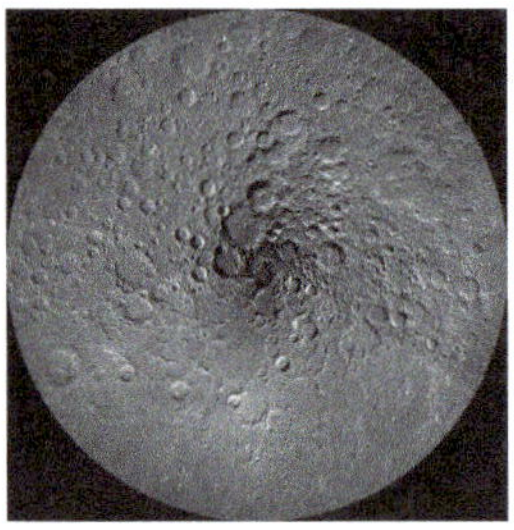

Lunar north pole

Lunar south pole

Moon – Oceanus Procellarum ("Ocean of Storms")

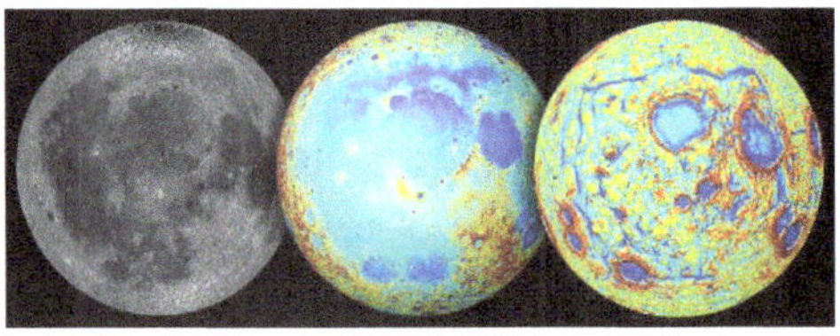

Ancient rift valleys – rectangular structure (visible – topography – GRAIL gravity gradients) (October 1, 2014).

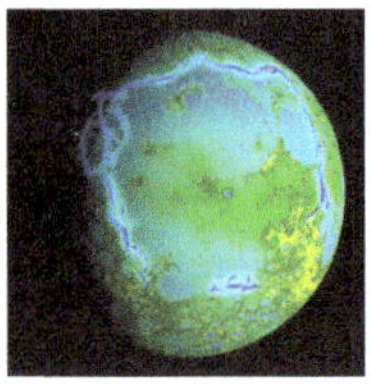

Ancient rift valleys – context.

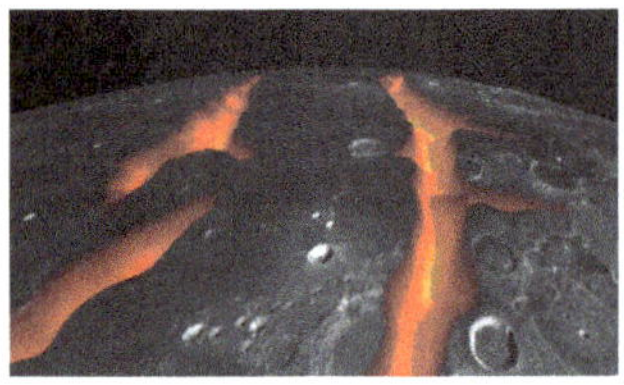

Ancient rift valleys – closeup (artist's concept).

Ocean Surface Topography

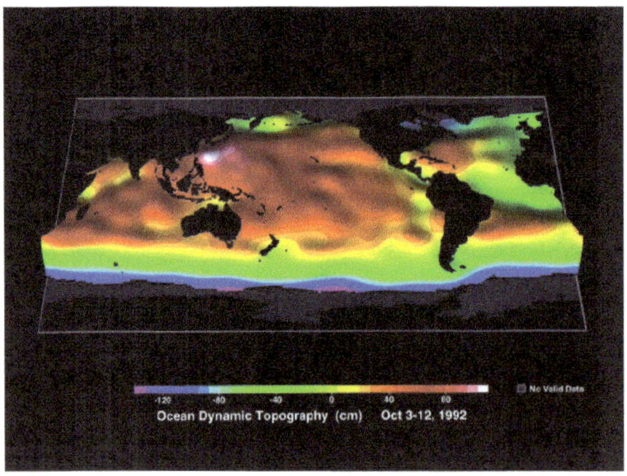

TOPEX/Poseidon was the first space mission that allowed scientists to map ocean topography with sufficient accuracy to study the large-scale current systems of the world's ocean. Although this image was constructed from only 10 days of TOPEX/ Poseidon data (October 3 to October 12, 1992), it reveals most of the current systems that have been identified by shipboard observations collected over the last 100 years.

The ocean surface has highs and lows, similar to the hills and valleys of Earth's land surface depicted on a topographic map. These variations, called ocean surface topography (or sea surface topography), also dynamic topography, are mapped using measurements of sea surface height relative to Earth's geoid. Earth's geoid is a calculated surface of equal gravitational potential energy and represents the shape the sea surface would be if the ocean were not in motion. Topography is the arrangement of natural and artificial features of an area. Ocean surface topography is specifically the distance between the height of the ocean surface from the geoid. Ocean surface topography is caused by ocean waves, tides, currents, and the loading of atmospheric pressure. The main purpose to measure ocean surface topography is to understand the large-scale circulation of the ocean. The height variations of ocean surface topography can be as much as two meters and are influenced by ocean circulation, ocean temperature, and salinity.

Ocean surface topography is used to map ocean currents, which move around the ocean's "hills" and "valleys" in predictable ways. A clockwise sense of rotation is found around "hills" in the northern hemisphere and "valleys" in the southern hemisphere. This is because of the Coriolis effect. Conversely, a counterclockwise sense of rotation is found around "valleys" in the northern hemisphere and "hills" in the southern hemisphere.

Ocean surface topography is also used to understand how the ocean moves heat around the globe, a critical component of Earth's climate, and for monitoring changes in global sea level. The sea surface height (SSH) is calculated through altimetry satellites, which determine the distance from the satellite to a target surface by measuring the satellite-to-surface round-trip time of a radar pulse. The satellites then measure the distance between their orbit altitude and the surface of the water. Due to the differing depths of the ocean, an approximation is made. This is called the Arbitrary Reference Surface. Arbitrary Reference Surface is an estimated surface that is calculated to factor in the shape of the Earth. The general shape of the earth is spherical, but flattened out at the North and South Pole. This approximated surface is called the reference ellipsoid. This enables data to be taken precisely due to the uniform surface level. The satellite's altitude then has to be calculated with respect to the reference ellipsoid. It is calculated using the orbital parameters of the satellite and various positioning instruments. The sea surface height is then the difference between the satellite's altitude relative to the reference ellipsoid and the altimeter range. The satellite sends microwave pulses to the ocean surface. The travel time of the pulses ascending to the oceans surface and back provides data of the sea surface height. In the image below you can see the measurement system using by the satellite Jason-1.

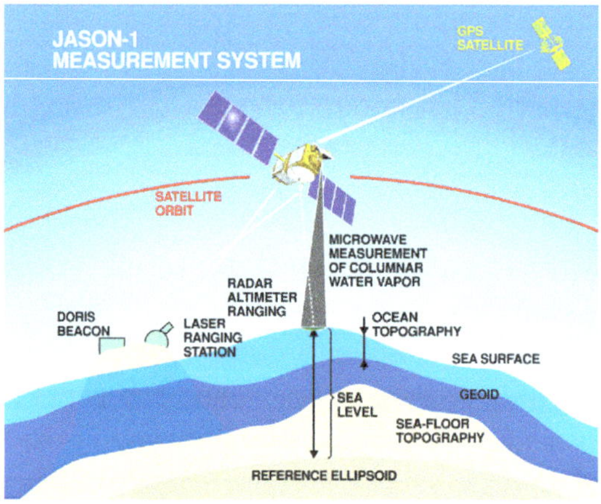

Jason-1 maps global ocean surface topography every 10 days.

Data Collection Purposes

The collection of the data is useful for the long-term information about the ocean and its currents. According to NASA science this data can also be used to provide understanding of weather, climate, navigation, fisheries management, and offshore operations. Observations made about the data are used to study the oceans tides, circulation, and the amount of heat the ocean contains. These observations can help predict short and long term effects of the weather and the earth's climate over time.

Satellites

Currently there are two different satellites calculating the earth ocean topography, Jason-1, and Jason-2. These two satellites are currently both in space orbiting Earth in a tandem rotation. They are approximately 330 kilometers apart.

Ocean surface topography can be derived from ship-going measurements of temperature and salinity at depth. However, since 1992, a series of satellite altimetry missions, beginning with TOPEX/Poseidon and continued with Jason-1 and the Ocean Surface Topography Mission on the Jason-2 satellite have measured sea surface height directly. By combining these measurements with gravity measurements from NASA's Grace mission, scientists can determine sea surface topography to within a few centimeters.

Jason-1 was launched by a Boeing Delta II rocket in California in 2001 and continued measurements initially collected by TOPEX/Poseidon satellite, which orbited from 1992 up until 2006. NASA and CNES, a French space agency, are joint partners in this mission.

The main objectives of the Jason-1 satellite is to collect data on the average ocean circulation around the globe in order to better understand its interaction with the time varying components and the involved mechanisms for initializing ocean models. To monitor the low frequency ocean variability and observe the season cycles and inter-annual variations like El Nino and La Nina, the North Atlantic oscillation, the pacific decadal oscillation, and planetary waves crossing the oceans over a period of months, then they will be modeled over a long period of time due to the precise altimetric observations. It aims to contribute to observations of the mesoscale ocean variability, affecting the whole oceans. This activity is especially intense near western boundary currents. Also monitor the average sea level because it is a large indicator of global warming through the sea level data. Improvement of tide modeling by observing more long period components such as costal interactions, internal waves, and tidal energy dissipation. Finally the satellite data will supply knowledge to support other types of marine meteorology which is the scientific study of the atmosphere.

Jason-2 was launched on June 20, 2008 by a Delta-2 rocket out of the California site in Vandenberg. Its current mission is to transfer the current data into a second decade of data collection on the ocean height. The long-term objectives of the Jason-2 satellite were to provide global descriptions of the seasonal and yearly changes of the circulation and heat storage in the ocean. This includes the study of short-term climatic changes such as El Nino, La Nina. The satellites detect global sea level mean and record the fluctuations. Also detecting the slow change of upper ocean circulation on decadal time scales, every ten years. Studying the transportation of heat and carbon in the ocean and examining the main components that fuel deep water tides. The satellites data collection also helps improve wind speed and height measurements in current time and for long term studies. Lastly improving our knowledge about the marine geoid. The first seven months Jason-2 was put into use it was flown in extreme close proximity to Jason-1. Only being one minute apart from each other the satellites observed the same area of the ocean. The reason for the close proximity in observation was for cross-calibration. This was meant to calculate any bias in the two altimeters. This multiple month observation proved that there was no bias in the data and both collections of data were consistent.

A new satellite mission called the Surface Water Ocean Topography Mission has been proposed to make the first global survey of the topography of all of Earth's surface water—the ocean, lakes and rivers. This study is aimed to provide a comprehensive view of Earth's freshwater bodies from space and more much detailed measurements of the ocean surface than ever before.

Organizations

The Ocean Surface Topography Mission is a combined effort from four different organizations with one mutual focus; measure sea surface height. The four contributing organizations are National Aeronautics and Space Administration (NASA), France's Centre National d'Etudes Spatiales (CNES), National Oceanographic and Atmospheric Administration (NOAA), and European Meteorological Satellite Organisation (EUMETSAT).

Terrain

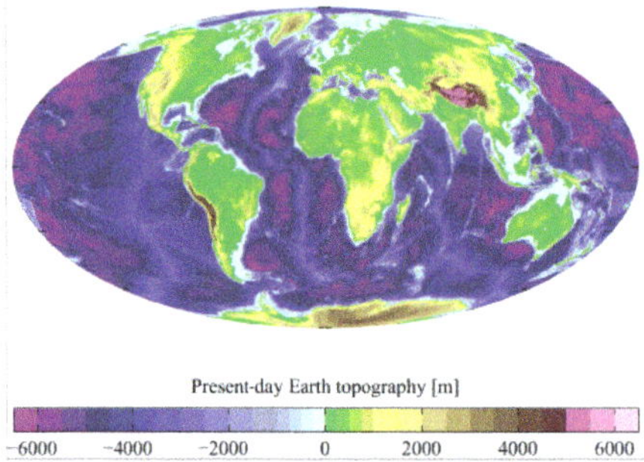

Present-day Earth altimetry and bathymetry. Data from the National Geophysical Data Center's TerrainBase Digital Terrain Model.

A shaded and colored image (i.e. terrain is enhanced) of varied terrain from the Shuttle Radar Topography Mission. This shows elevation model of New Zealand's Alpine Fault running about 500 km (300 mi) long. The escarpment is flanked by a vast chain of hills between the fault and the mountains of New Zealand's Southern Alps. Northeast is towards the top.

Terrain or relief (also topographical relief) is the vertical and horizontal dimension of the land surface. When relief is described underwater, the term bathymetry is used. The Latin Word "Terra (The root word of terrain)," is "Earth."

Terrain is used as a general term in physical geography, referring to the lay of the land. This is usually expressed in terms of the elevation, slope, and orientation of terrain features. Terrain affects surface water flow and distribution. Over a large area, it can affect weather and climate patterns.

Importance

The understanding of terrain is critical for many reasons:

- The terrain of a region largely determines its suitability for human settlement: flatter, alluvial plains tend to have better farming soils than steeper, rockier uplands.
- In terms of environmental quality, agriculture, and hydrology, understanding the terrain of an area enables the understanding of watershed boundaries, drainage characteristics, water movement, and impacts on water quality. Complex arrays of relief data are used as input parameters for hydrology transport models (such as the SWMM or DSSAM Models) to allow prediction of river water quality.
- Understanding terrain also supports soil conservation, especially in agriculture. Contour ploughing is an established practice enabling sustainable agriculture on sloping land; it is the practice of ploughing along lines of equal elevation instead of up and down a slope.
- Terrain is militarily critical because it determines the ability of armed forces to take and hold areas, and move troops and material into and through areas. An understanding of terrain is basic to both defensive and offensive strategy.
- Terrain is important in determining weather patterns. Two areas geographically close to each other may differ radically in precipitation levels or timing because of elevation differences or a “rain shadow” effect.
- Precise knowledge of terrain is vital in aviation, especially for low-flying routes and maneuvers and airport altitudes. Terrain will also affect range and performance of radars and terrestrial radio navigation systems. Furthermore, a hilly or mountainous terrain can strongly impact the implementation of a new aerodrome and the orientation of its runways.

Relief

Relief (or *local relief*) refers specifically to the quantitative measurement of vertical elevation change in a landscape. It is the difference between maximum and minimum elevations within a given area, usually of limited extent. The relief of a landscape can change with the size of the area over which it is measured, making the definition of the scale over which it is measured very important. Because it is related to the slope of surfaces within the area of interest and to the gradient of any streams present, the relief of a landscape is a useful metric in the study of the Earth’s surface. Relief energy, which may be defined *inter alia* as "the maximum height range in a regular grid", is essentially an indication of the ruggedness or relative height of the terrain.

Geomorphology

Geomorphology is in large part the study of the formation of terrain or topography. Terrain is formed by intersecting processes:

- Geological processes: Migration of tectonic plates, faulting and folding, volcanic eruptions, rivers.
- Erosional processes: water and wind erosion, landslides.
- Extraterrestrial: meteorite impacts.

Tectonic processes such as orogenies cause land to be elevated, and erosional or weathering processes cause land to be worn away to lower elevations.

Land surface parameters are quantitative measures of various morphometric properties of a surface. The most common examples are used to derive slope or aspect of a terrain or curvatures at each location. These measures can also be used to derive hydrological parameters that reflect flow/erosion processes. Climatic parameters are based on the modelling of solar radiation or air flow.

Land surface objects, or landforms, are definite physical objects (lines, points, areas) that differ from the surrounding objects. The most typical examples airlines of watersheds, stream patterns, ridges, break-lines, pools or borders of specific landforms.

Elevation

A sign at 8000 feet (2438 meters) in the San Bernardino Mountains.

The elevation of a geographic location is its height above or below a fixed reference point, most commonly a reference geoid, a mathematical model of the Earth's sea level as an equipotential gravitational surface (see Geodetic system, vertical datum). Elevation, or geometric height, is mainly used when referring to points on the Earth's surface, while altitude or geopotential height is used for points above the surface, such as an aircraft in flight or a spacecraft in orbit, and depth is used for points below the surface.

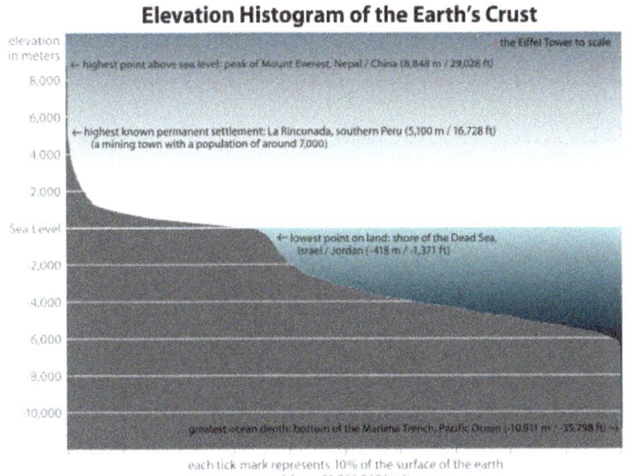

Elevation histogram of the surface of the Earth – approximately 71% of the Earth's surface is covered with water.

Elevation is not to be confused with the distance from the center of the Earth; due to equatorial bulge, the summits of Mt. Everest and Chimborazo have, respectively, the largest elevation and the largest geocentric distance.

Maps and GIS

A topographical map is the main type of map used to depict elevation, often through use of contour lines. In a Geographic Information System (GIS), digital elevation models (DEM) are commonly used to represent the surface (topography) of a place, through a raster (grid) dataset of elevations. Digital terrain models are another way to represent terrain in GIS.

Part of a topographic map of Haleakala (Hawaii), showing elevation.

Landsat Image over SRTM Elevation by NASA, showing the Cape Peninsula and Cape of Good Hope, South Africa in the foreground.

Heightmap of Earth's surface (including water and ice) in equirectangular projection, normalized as 8-bit grayscale, where lighter values indicate higher elevation.

Global 1-kilometer Map

This map is derived from GTOPO30 data that describes the elevation of Earth's terrain at intervals of 30 arcseconds (approximately 1 km). It uses color and shading instead of contour lines to indicate elevation.

Hypsography

Hypsography is the study of the distribution of elevations on the surface of the Earth, although the term is sometimes also applied to other rocky planets such as Mars or Venus. The term originates from the Greek word □ψος "hypsos" meaning height. Most often it is used only in reference to elevation of land but a complete description of Earth's solid surface requires a description of the seafloor as well. Related to the term hypsometry, the measurement of these elevations of a planet's solid surface are taken relative to mean datum, except for Earth which is taken relative to the sea level.

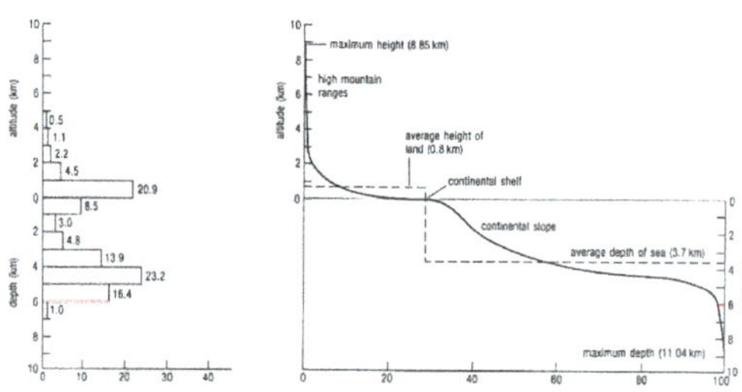

Hypsography of the Earth. Notice that Earth has two peaks in elevation, one for the continents, the other for the ocean floors.

Temperature

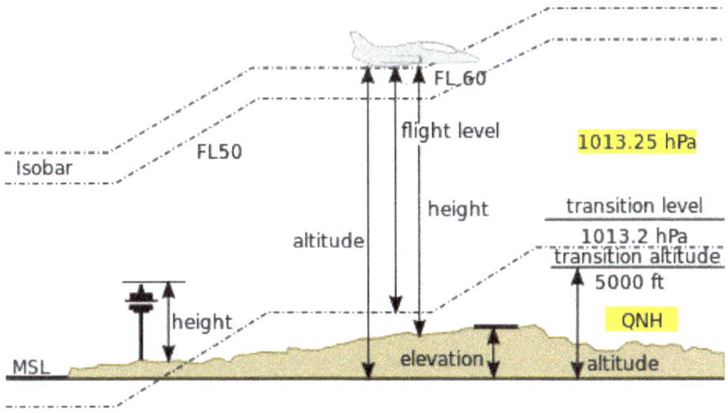

Vertical Distance Comparison

In the troposphere, temperatures decrease with altitude. This lapse rate is approximately 6.5 °C/km.

Geomorphology

Badlands incised into shale at the foot of the North Caineville Plateau, Utah, within the pass carved by the Fremont River and known as the Blue Gate. GK Gilbert studied the landscapes of this area in great detail, forming the observational foundation for many of his studies on geomorphology.

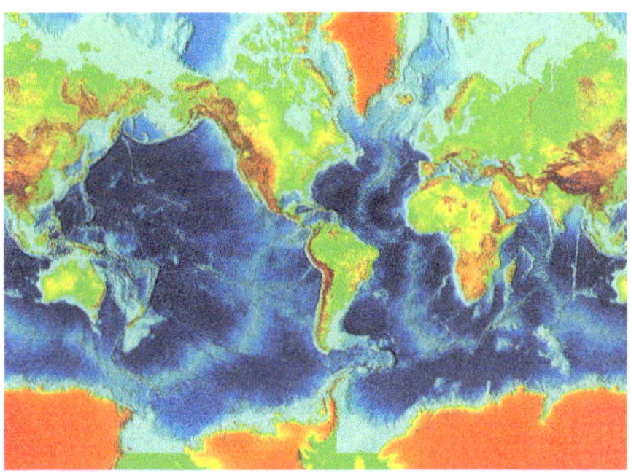

Surface of the Earth, showing higher elevations in red.

Geomorphology (from Ancient Greek: γ□, *gê*, "earth"; μορφή, *morph*□, "form"; and λόγος, *lógos*, "study") is the scientific study of the origin and evolution of topographic and bathymetric features created by physical, chemical or biological processes operating at or near the Earth's surface. Geomorphologists seek to understand why landscapes look the way they do, to understand landform history and dynamics and to predict changes through a combination of field observations, physical experiments and numerical modeling. Geomorphologists work within disciplines such as physical geography, geology, geodesy, engineering geology, archaeology and geotechnical engineering. This broad base of interests contributes to many research styles and interests within the field.

Overview

Waves and water chemistry lead to structural failure in exposed rocks

Earth's surface is modified by a combination of surface processes that sculpt landscapes, and geologic processes that cause tectonic uplift and subsidence, and shape the coastal geography. Surface processes comprise the action of water, wind, ice, fire, and living things on the surface of the Earth, along with chemical reactions that form soils and alter material properties, the stability and rate of change of topography under the force of gravity, and other factors, such as (in the very recent past) human alteration of the landscape. Many of these factors are strongly mediated by climate. Geologic processes include the uplift of mountain ranges, the growth of volcanoes, isostatic changes in land surface elevation (sometimes in response to surface processes), and the formation of deep sedimentary basins where the surface of the Earth drops and is filled with material eroded from other parts of the landscape. The Earth's surface and its topography therefore are an intersection of climatic, hydrologic, and biologic action with geologic processes, or alternatively stated, the intersection of the Earth's lithosphere with its hydrosphere, atmosphere, and biosphere.

The broad-scale topographies of the Earth illustrate this intersection of surface and subsurface action. Mountain belts are uplifted due to geologic processes. Denudation of these high uplifted regions produces sediment that is transported and deposited elsewhere within the landscape or off the coast. On progressively smaller scales, similar ideas apply, where individual landforms evolve in response to the balance of additive processes (uplift and deposition) and subtractive processes (subsidence and erosion). Often, these processes directly affect each other: ice sheets, water, and sediment are all loads that change topography through flexural isostasy. Topography can modify the local climate, for example through orographic precipitation, which in turn modifies the topography by changing the hydrologic regime in which it evolves. Many geomorphologists are particularly interested in the potential for feedbacks between climate and tectonics, mediated by geomorphic processes.

In addition to these broad-scale questions, geomorphologists address issues that are more specific and/or more local. Glacial geomorphologists investigate glacial deposits such as moraines, eskers, and proglacial lakes, as well as glacial erosional features, to build chronologies of both small glaciers and large ice sheets and understand their motions and effects upon the landscape. Fluvial geomorphologists focus on rivers, how they transport sediment, migrate across the landscape, cut into bedrock, respond to environmental and tectonic changes, and interact with humans. Soils geomorphologists investigate soil profiles and chemistry to learn about the history of a particular landscape and understand how climate, biota, and rock interact. Other geomorphologists study how hillslopes form and change. Still others investigate the relationships between ecology and geomorphology. Because geomorphology is defined to comprise everything related to the surface of the Earth and its modification, it is a broad field with many facets.

Geomorphologists use a wide range of techniques in their work. These may include fieldwork and field data collection, the interpretation of remotely sensed data, geochemical analyses, and the numerical modelling of the physics of landscapes. Geomorphologists may rely on geochronology, using dating methods to measure the rate of changes to the surface. Terrain measurement techniques are vital to quantitatively describe the form of the Earth's surface, and include differential GPS, remotely sensed digital terrain models and laser scanning, to quantify, study, and to generate illustrations and maps.

Practical applications of geomorphology include hazard assessment (such as landslide prediction and mitigation), river control and stream restoration, and coastal protection. Planetary geomorphology studies landforms on other terrestrial planets such as Mars. Indications of effects of wind, fluvial, glacial, mass wasting, meteor impact, tectonics and volcanic processes are studied. This effort not only helps better understand the geologic and atmospheric history of those planets but also extends geomorphological study of the Earth. Planetary geomorphologists often use Earth analogues to aid in their study of surfaces of other planets.

History

Other than some notable exceptions in antiquity, geomorphology is a relatively young science, growing along with interest in other aspects of the earth sciences in the mid-

19th century. This section provides a very brief outline of some of the major figures and events in its development.

"Cono de Arita" at the dry lake Salar de Arizaro on the Atacama Plateau, in northwestern Argentina. The cone itself is a volcanic edifice, representing complex interaction of intrusive igneous rocks with the surrounding salt.

Lake "Veľké Hincovo pleso" in High Tatras, Slovakia. The lake occupies an "overdeepening" carved by flowing ice that once occupied this glacial valley.

Ancient Geomorphology

The first theory of geomorphology was arguably devised by the polymath Chinese scientist and statesman Shen Kuo (1031–1095 AD). This was based on his observation of marine fossil shells in a geological stratum of a mountain hundreds of miles from the Pacific Ocean. Noticing bivalve shells running in a horizontal span along the cut section of a cliffside, he theorized that the cliff was once the pre-historic location of a seashore that had shifted hundreds of miles over the centuries. He inferred that the land was reshaped and formed by soil erosion of the mountains and by deposition of silt, after observing strange natural erosions of the Taihang Mountains and the Yandang Mountain near Wenzhou. Furthermore, he promoted the theory of gradual climate change over centuries of time once ancient petrified bamboos were found to be preserved underground in the dry, northern climate zone of *Yanzhou*, which is now modern day Yan'an, Shaanxi province.

Early Modern Geomorphology

The term geomorphology seems to have been first used by Laumann in an 1858 work written in German. Keith Tinkler has suggested that the word came into general use in

English, German and French after John Wesley Powell and W. J. McGee used it during the International Geological Conference of 1891. John Edward Marr in his The Scientific Study of Scenery considered his book as, 'an Introductory Treatise on Geomorphology, a subject which has sprung from the union of Geology and Geography'.

An early popular geomorphic model was the *geographical cycle* or *cycle of erosion* model of broad-scale landscape evolution developed by William Morris Davis between 1884 and 1899. It was an elaboration of the uniformitarianism theory that had first been proposed by James Hutton (1726–1797). With regard to valley forms, for example, uniformitarianism posited a sequence in which a river runs through a flat terrain, gradually carving an increasingly deep valley, until the side valleys eventually erode, flattening the terrain again, though at a lower elevation. It was thought that tectonic uplift could then start the cycle over. In the decades following Davis's development of this idea, many of those studying geomorphology sought to fit their findings into this framework, known today as "Davisian". Davis's ideas are of historical importance, but have been largely superseded today, mainly due to their lack of predictive power and qualitative nature.

In the 1920s, Walther Penck developed an alternative model to Davis's. Penck thought that landform evolution was better described as an alternation between ongoing processes of uplift and denudation, as opposed to Davis's model of a single uplift followed by decay. He also emphasised that in many landscapes slope evolution occurs by backwearing of rocks, not by Davisian-style surface lowering, and his science tended to emphasise surface process over understanding in detail the surface history of a given locality. Penck was German, and during his lifetime his ideas were at times rejected vigorously by the English-speaking geomorphology community. His early death, Davis' dislike for his work, and his at-times-confusing writing style likely all contributed to this rejection.

Both Davis and Penck were trying to place the study of the evolution of the Earth's surface on a more generalized, globally relevant footing than it had been previously. In the early 19th century, authors – especially in Europe – had tended to attribute the form of landscapes to local climate, and in particular to the specific effects of glaciation and periglacial processes. In contrast, both Davis and Penck were seeking to emphasize the importance of evolution of landscapes through time and the generality of the Earth's surface processes across different landscapes under different conditions.

During the early 1900s, the study of regional-scale geomorphology was termed "physiography". Physiography later was considered to be a contraction of "physical" and "geography", and therefore synonymous with physical geography, and the concept became embroiled in controversy surrounding the appropriate concerns of that discipline. Some geomorphologists held to a geological basis for physiography and emphasized a concept of physiographic regions while a conflicting trend among geographers was to equate physiography with "pure morphology," separated from its geological heritage.

In the period following World War II, the emergence of process, climatic, and quantitative studies led to a preference by many earth scientists for the term "geomorphology" in order to suggest an analytical approach to landscapes rather than a descriptive one.

Quantitative Geomorphology

Part of the Great Escarpment in the Drakensberg, southern Africa. This landscape, with its high altitude plateau being incised into by the steep slopes of the escarpment, was cited by Davis as a classic example of his cycle of erosion.

Geomorphology was started to be put on a solid quantitative footing in the middle of the 20th century. Following the early work of Grove Karl Gilbert around the turn of the 20th century, a group of natural scientists, geologists and hydraulic engineers including Ralph Alger Bagnold, Hans-Albert Einstein, Frank Ahnert, John Hack, Luna Leopold, A. Shields, Thomas Maddock, Arthur Strahler, Stanley Schumm, and Ronald Shreve began to research the form of landscape elements such as rivers and hillslopes by taking systematic, direct, quantitative measurements of aspects of them and investigating the scaling of these measurements. These methods began to allow prediction of the past and future behavior of landscapes from present observations, and were later to develop into the modern trend of a highly quantitative approach to geomorphic problems. Quantitative geomorphology can involve fluid dynamics and solid mechanics, geomorphometry, laboratory studies, field measurements, theoretical work, and full landscape evolution modeling. These approaches are used to understand weathering and the formation of soils, sediment transport, landscape change, and the interactions between climate, tectonics, erosion, and deposition.

Contemporary Geomorphology

Today, the field of geomorphology encompasses a very wide range of different approaches and interests. Modern researchers aim to draw out quantitative "laws" that govern Earth surface processes, but equally, recognize the uniqueness of each landscape and environment in which these processes operate. Particularly important realizations in contemporary geomorphology include:

1) that not all landscapes can be considered as either "stable" or "perturbed", where this perturbed state is a temporary displacement away from some ideal target form.

Instead, dynamic changes of the landscape are now seen as an essential part of their nature.

2) that many geomorphic systems are best understood in terms of the stochasticity of the processes occurring in them, that is, the probability distributions of event magnitudes and return times. This in turn has indicated the importance of chaotic determinism to landscapes, and that landscape properties are best considered statistically. The same processes in the same landscapes do not always lead to the same end results.

Processes

Gorge cut by the Indus river into bedrock, Nanga Parbat region, Pakistan. This is the deepest river canyon in the world. Nanga Parbat itself, the world's 9th highest mountain, is seen in the background.

Geomorphically relevant processes generally fall into (1) the production of regolith by weathering and erosion, (2) the transport of that material, and (3) its eventual deposition. Primary surface processes responsible for most topographic features include wind, waves, chemical dissolution, mass wasting, groundwater movement, surface water flow, glacial action, tectonism, and volcanism. Other more exotic geomorphic processes might include periglacial (freeze-thaw) processes, salt-mediated action, marine currents activity, seepage of fluids through the seafloor or extraterrestrial impact.

Aeolian Processes

Wind-eroded alcove near Moab, Utah

Aeolian processes pertain to the activity of the winds and more specifically, to the winds' ability to shape the surface of the Earth. Winds may erode, transport, and deposit materials, and are effective agents in regions with sparse vegetation and a large supply of fine, unconsolidated sediments. Although water and mass flow tend to mobilize more material than wind in most environments, eolian processes are important in arid environments such as deserts.

Biological Processes

Beaver dams, as this one in Tierra del Fuego, constitute a specific form of zoogeomorphology, a type of biogeomorphology.

The interaction of living organisms with landforms, or biogeomorphologic processes, can be of many different forms, and is probably of profound importance for the terrestrial geomorphic system as a whole. Biology can influence very many geomorphic processes, ranging from biogeochemical processes controlling chemical weathering, to the influence of mechanical processes like burrowing and tree throw on soil development, to even controlling global erosion rates through modulation of climate through carbon dioxide balance. Terrestrial landscapes in which the role of biology in mediating surface processes can be definitively excluded are extremely rare, but may hold important information for understanding the geomorphology of other planets, such as Mars.

Fluvial Processes

Seif and barchan dunes in the Hellespontus region on the surface of Mars. Dunes are mobile landforms created by the transport of large volumes of sand by wind.

Rivers and streams are not only conduits of water, but also of sediment. The water, as it flows over the channel bed, is able to mobilize sediment and transport it downstream, either as bed load, suspended load or dissolved load. The rate of sediment transport depends on the availability of sediment itself and on the river's discharge. Rivers are also capable of eroding into rock and creating new sediment, both from their own beds and also by coupling to the surrounding hillslopes. In this way, rivers are thought of as setting the base level for large scale landscape evolution in nonglacial environments. Rivers are key links in the connectivity of different landscape elements.

As rivers flow across the landscape, they generally increase in size, merging with other rivers. The network of rivers thus formed is a drainage system. These systems take on four general patterns, dendritic, radial, rectangular, and trellis. Dendritic happens to be the most common occurring when the underlying strata is stable (without faulting). Drainage systems have four primary components: drainage basin, alluvial valley, delta plain, and receiving basin. Some geomorphic examples of fluvial landforms are alluvial fans, oxbow lakes, and fluvial terraces.

Glacial Processes

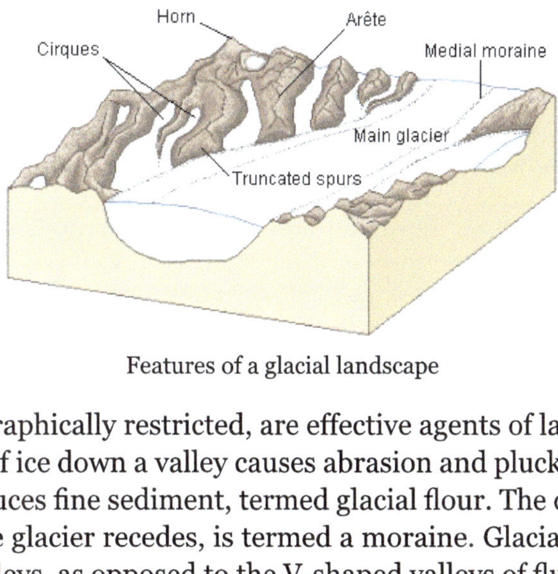

Features of a glacial landscape

Glaciers, while geographically restricted, are effective agents of landscape change. The gradual movement of ice down a valley causes abrasion and plucking of the underlying rock. Abrasion produces fine sediment, termed glacial flour. The debris transported by the glacier, when the glacier recedes, is termed a moraine. Glacial erosion is responsible for U-shaped valleys, as opposed to the V-shaped valleys of fluvial origin.

The way glacial processes interact with other landscape elements, particularly hillslope and fluvial processes, is an important aspect of Plio-Pleistocene landscape evolution and its sedimentary record in many high mountain environments. Environments that have been relatively recently glaciated but are no longer may still show elevated landscape change rates compared to those that have never been glaciated. Nonglacial geomorphic processes which nevertheless have been conditioned by past glaciation are termed paraglacial processes. This concept contrasts with periglacial processes, which are directly driven by formation or melting of ice or frost.

Hillslope Processes

Talus cones on the north shore of Isfjorden, Svalbard, Norway. Talus cones are accumulations of coarse hillslope debris at the foot of the slopes producing the material.

The Ferguson Slide is an active landslide in the Merced River canyon on California State Highway 140, a primary access road to Yosemite National Park.

Soil, regolith, and rock move downslope under the force of gravity via creep, slides, flows, topples, and falls. Such mass wasting occurs on both terrestrial and submarine slopes, and has been observed on Earth, Mars, Venus, Titan and Iapetus.

Ongoing hillslope processes can change the topology of the hillslope surface, which in turn can change the rates of those processes. Hillslopes that steepen up to certain critical thresholds are capable of shedding extremely large volumes of material very quickly, making hillslope processes an extremely important element of landscapes in tectonically active areas.

On the Earth, biological processes such as burrowing or tree throw may play important roles in setting the rates of some hillslope processes.

Igneous Processes

Both volcanic (eruptive) and plutonic (intrusive) igneous processes can have important impacts on geomorphology. The action of volcanoes tends to rejuvenize landscapes, covering the old land surface with lava and tephra, releasing pyroclastic material and forcing rivers through new paths. The cones built by eruptions also build substantial new topography, which can be acted upon by other surface processes. Plutonic rocks intruding then solidifying at depth can cause both uplift or subsidence of the surface, depending on whether the new material is denser or less dense than the rock it displaces.

Tectonic Processes

Tectonic effects on geomorphology can range from scales of millions of years to minutes or less. The effects of tectonics on landscape are heavily dependent on the nature of the underlying bedrock fabric that more or less controls what kind of local morphology tectonics can shape. Earthquakes can, in terms of minutes, submerge large areas of land creating new wetlands. Isostatic rebound can account for significant changes over hundreds to thousands of years, and allows erosion of a mountain belt to promote further erosion as mass is removed from the chain and the belt uplifts. Long-term plate tectonic dynamics give rise to orogenic belts, large mountain chains with typical lifetimes of many tens of millions of years, which form focal points for high rates of fluvial and hillslope processes and thus long-term sediment production.

Features of deeper mantle dynamics such as plumes and delamination of the lower lithosphere have also been hypothesised to play important roles in the long term (> million year), large scale (thousands of km) evolution of the Earth's topography. Both can promote surface uplift through isostasy as hotter, less dense, mantle rocks displace cooler, denser, mantle rocks at depth in the Earth.

Marine Processes

Marine processes are those associated with the action of waves, marine currents and seepage of fluids through the seafloor.

Scales in Geomorphology

Different geomorphological processes dominate at different spatial and temporal scales. Moreover, scales on which processes occur may determine the reactivity or otherwise of landscapes to changes in driving forces such as climate or tectonics. These ideas are key to the study of geomorphology today.

To help categorize landscape scales some geomorphologists might use the following taxonomy:

- 1st – Continent, ocean basin, climatic zone (~10,000,000 km^2)
- 2nd – Shield, e.g. Baltic Shield, or mountain range (~1,000,000 km^2)
- 3rd – Isolated sea, Sahel (~100,000 km^2)
- 4th – Massif, e.g. Massif Central or Group of related landforms, e.g., Weald (~10,000 km^2)
- 5th – River valley, Cotswolds (~1,000 km^2)

- 6th – Individual mountain or volcano, small valleys (~100 km²)
- 7th – Hillslopes, stream channels, estuary (~10 km²)
- 8th – gully, barchannel (~1 km²)
- 9th – Meter-sized features

Overlap with other Fields

There is a considerable overlap between geomorphology and other fields. Deposition of material is extremely important in sedimentology. Weathering is the chemical and physical disruption of earth materials in place on exposure to atmospheric or near surface agents, and is typically studied by soil scientists and environmental chemists, but is an essential component of geomorphology because it is what provides the material that can be moved in the first place. Civil and environmental engineers are concerned with erosion and sediment transport, especially related to canals, slope stability (and natural hazards), water quality, coastal environmental management, transport of contaminants, and stream restoration. Glaciers can cause extensive erosion and deposition in a short period of time, making them extremely important entities in the high latitudes and meaning that they set the conditions in the headwaters of mountain-born streams; glaciology therefore is important in geomorphology.

Topographic Prominence

In topography, prominence characterizes the height of a mountain or hill's summit by the vertical distance between it and the lowest contour line encircling it but containing no higher summit within it. It is a measure of the independence of a summit. A peak's key col is a unique point on this contour line and the parent peak is some higher mountain, selected according to various objective criteria.

Definitions

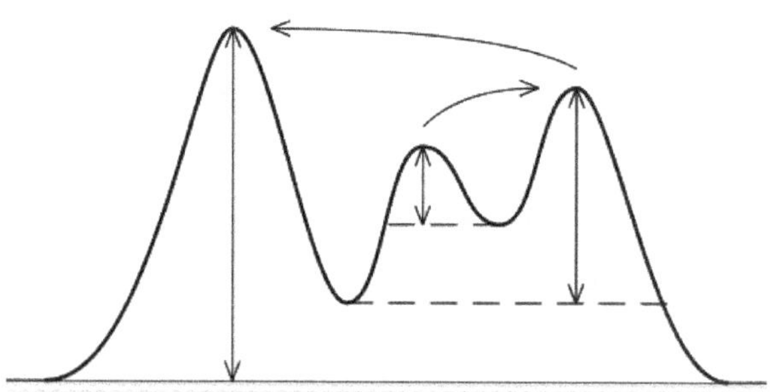

Figure 1. Vertical arrows show the topographic prominence of three peaks on an island. The dashed horizontal lines show the lowest contours that do not encircle higher peaks. Curved arrows point from a peak to its parent.

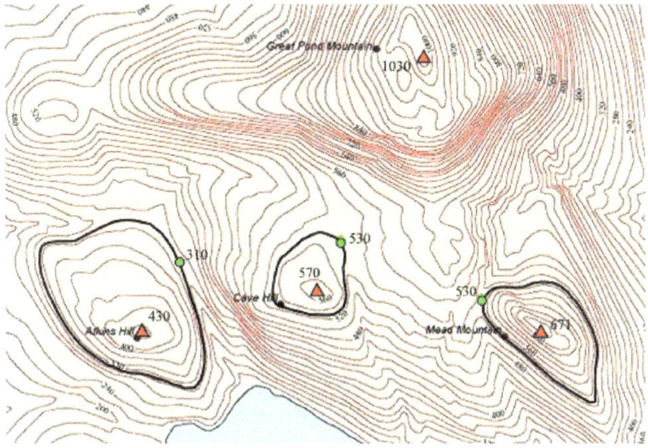

Topographic prominence of three peaks near Great Pond Mountain, Maine, USA. Red triangles mark the peaks, the lowest contour line encircling each peak are shown in black and the green dots mark the key cols. The prominences are Atkins Hill: 430 – 310 = 120 ft, Cave Hill: 570 – 530 = 40 ft, Mead Mountain: 671 – 530 = 141 ft. The parent peak of each peak is Great Pond Mountain.

By convention, the prominence of Mount Everest, the Earth's highest mountain, is taken to equal the elevation of its summit above sea level. Apart from this special case, there are several equivalent definitions:

- The prominence of a peak is the height of the peak's summit above the lowest contour line encircling it but containing no higher summit within it.
- If the peak's prominence is *P* metres, to get from the summit to any higher terrain one must descend at least *P* metres. Together with the convention for Mount Everest, this implies that the prominence of any island or continental highpoint is equal to its elevation above sea level.
- For every ridge (or path of any kind) connecting the peak to higher terrain, find the lowest point on the ridge. This will be at a col (also called a saddle point or pass). The *key col* (or *key saddle*, or *linking col*, or *link*) is defined as the highest of these cols, along all connecting ridges. (If the peak is the highest point on a landmass, the key col will be the ocean, and the prominence of the peak is equal to its elevation.) The prominence is the difference between the elevation of the peak and the elevation of the key col. See Figure 1.
- Suppose that the sea level rises to the lowest level at which the peak becomes the highest point on an island. The prominence of that peak is the height of that island. The key col represents the last isthmus connecting the island to a higher island, just before they become disconnected.

In Mountaineering

Prominence is interesting to some mountaineers because it is an objective measurement that is strongly correlated with the subjective significance of a summit. Peaks with low prominences are either subsidiary tops of some higher summit or relatively insignificant independent summits. Peaks with high prominences tend to be the highest points around and are likely to have extraordinary views.

Only summits with a sufficient degree of prominence are regarded as independent mountains. For example, the world's second-highest mountain is K2 (height 8,611 m, prominence 4,017 m). While Mount Everest's South Summit (height 8,749 m, prominence about 10 m) is taller than K2, it is not considered an independent mountain because it is a subsummit of the main summit (which has a height and prominence of 8,848 m).

Many lists of mountains take topographic prominence as a criterion for inclusion, or *cutoff*. John and Anne Nuttall's *The Mountains of England and Wales* uses a cutoff of 15 m (about 50 ft), and Alan Dawson's list of Marilyns uses 150 m (about 500 ft). (Dawson's list and the term "Marilyn" are limited to Britain and Ireland.) In the contiguous United States, the famous list of "fourteeners" (14,000 foot / 4268 m peaks) uses a cutoff of 300 ft / 91 m (with some exceptions). Also in the U.S., 2000 feet (610 m) of prominence has become an informal threshold that signifies that a peak has major stature. Lists with a high topographic prominence cutoff tend to favour isolated peaks or those that are the highest point of their massif; a low value, such as the Nuttalls', results in a list with many summits that may be viewed by some as insignificant.

While the use of prominence as a cutoff to form a list of peaks ranked by elevation is standard, and is the most common use of the concept, it is also possible to use prominence as a mountain measure in itself. This generates lists of peaks ranked by prominence, which are qualitatively different from lists ranked by elevation. Such lists tend to emphasize isolated high peaks, such as range or island high points and stratovolcanoes. One advantage of a prominence-ranked list is that it needs no cutoff, since a peak with high prominence is automatically an independent peak.

Parent Peak

It is common to define a peak's parent as a particular peak in the higher terrain connected to the peak by the key col. If there are many higher peaks there are various ways of defining which one is the parent, not necessarily based on geological or geomorphological factors. The "parent" relationship defines a hierarchy which defines some peaks as subpeaks of others. For example, in Figure 1, the middle peak is a subpeak of the right peak, which is in turn a subpeak of the left peak, which is the highest point on its landmass. In that example, there is no controversy over the hierarchy; in practice, there are different definitions of parent. These different definitions follow.

A special case occurs for the highest point on an oceanic island or continent. Some sources define no parent in this case; others treat Mount Everest as the parent of every such peak with the ocean as the "key col".

Encirclement or Island Parentage

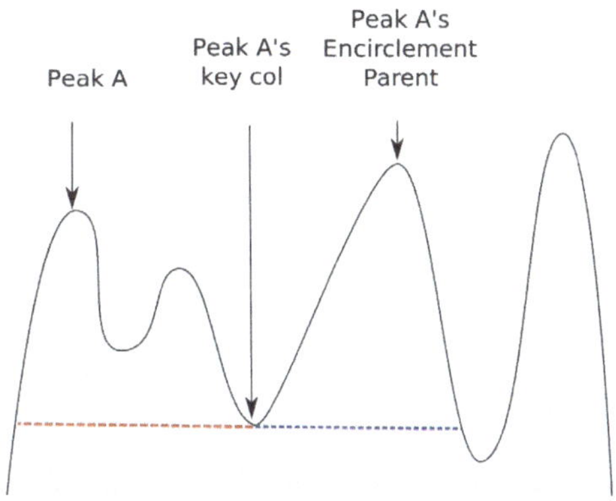

Figure 2. Showing two closed contour lines meeting at Peak A's key col.

Also called *prominence island parentage*, this is the most mathematically natural definition, and is defined as follows. In figure 2 the key col of peak A is at the meeting place of two closed contours, one encircling A (and no higher peaks) and the other containing at least one higher peak. The encirclement parent of A is the highest peak that is inside this other contour. In terms of the rising-sea model, the two contours together bound an island, with two pieces connected by an isthmus at the key col. The encirclement parent is the highest point on this entire island.

For example, the encirclement parent of Mont Blanc, the highest peak in the Alps, is Mount Everest. Mont Blanc's key col is a piece of low ground near Lake Onega in northwestern Russia (at 113 m elevation), on the divide between lands draining into the Baltic and Caspian Seas. This is the meeting place of two 113 m contours, one of them encircling Mont Blanc; the other contour encircles Mount Everest. This example demonstrates that the encirclement parent can be very far away from the peak in question when the key col is low.

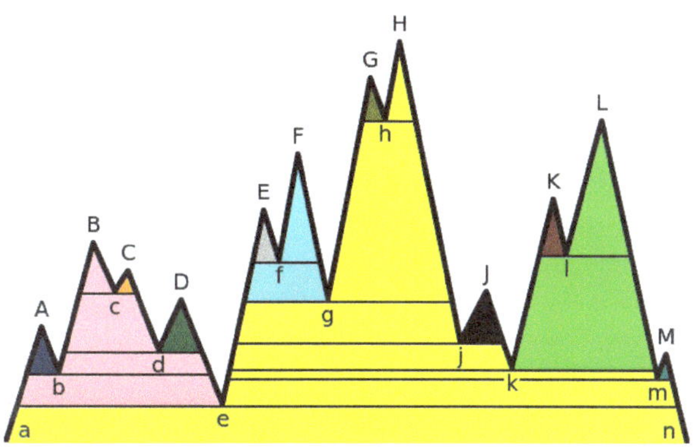

Figure 3. Diagram of a mountain range showing peaks and cols, from which mountain parentage and prominences can be determined.

This means that, while simple to define, the encirclement parent often does not satisfy the intuitive requirement that the parent peak should be close to the child peak. For example, one common use of the concept of parent is to make clear the location of a peak. If we say that Peak A has Mont Blanc for a parent, we would expect to find Peak A somewhere close to Mont Blanc. This is not always the case for the various concepts of parent, and is least likely to be the case for encirclement parentage.

Figure 3 shows a schematic range of peaks with the color underlying the minor peaks indicating the encirclement parent. In this case the encirclement parent of M is H whereas an intuitive view might be that L was the parent. Indeed, if col "k" were slightly lower, L would be the true encirclement parent.

The encirclement parent is the highest possible parent for a peak; all other definitions pick out a (possibly different) peak on the combined island, a "closer" peak than the encirclement parent (if there is one), which is still "better" than the peak in question. The differences lie in what criteria are used to define "closer" and "better."

Prominence Parentage

The (prominence) parent peak of peak A can be found by dividing the island or region in question into territories, by tracing the runoff from the key col of every peak that is more prominent than peak A. The parent is the peak whose territory peak A is in.

For hills with low prominence in Britain, a definition of 'parent Marilyn' is sometimes used to classify low hills. This is found by dividing the region of Britain in question into territories, *one for each Marilyn* ("Marilyn" being a British term for a hill with a prominence of at least 150 m). The parent Marilyn is the Marilyn whose territory the hill's summit is in. If the hill is on an island (in Britain) whose highest point is less than 150m, it has no parent Marilyn.

Prominence parentage is the only definition used in the British Isles because encirclement parentage breaks down when the key col approaches sea level. Using the encirclement definition, the parent of almost any small hill in a low-lying coastal area would be Ben Nevis, an unhelpful and confusing outcome. Meanwhile, 'height' parentage is not used because there is no obvious choice of cutoff.

This choice of method might at first seem arbitrary, but it provides every hill with a clear and unambiguous parent peak that is taller and more prominent than the hill itself, while also being connected to it (via ridge lines). The parent of a low hill will also usually be nearby; this becomes less likely as the hill's height and prominence increase. Using prominence parentage, one may produce a 'hierarchy' of peaks going back to the highest point on the island. One such chain in Britain would read:

Billinge Hill → Winter Hill → Hail Storm Hill → Boulsworth Hill → Kinder Scout → Cross Fell → Helvellyn → Scafell Pike → Snowdon → Ben Nevis.

At each stage in the chain, both height and prominence increase.

Line Parentage

Line parentage, also called height parentage, is similar to prominence parentage, but it requires a prominence cutoff criterion. The height parent is the closest peak to peak A (along all ridges connected to A) that has a greater height than A, and satisfies some prominence criteria.

The disadvantage of this concept is that it goes against the intuition that a parent peak should always be more significant than its child. However it can be used to build an entire lineage for a peak which contains a great deal of information about the peak's position.

Other Criteria

To choose among possible parents, instead of choosing the closest possible parent, it is possible to choose the one which requires the least descent along the ridge.

In general, the analysis of parents and lineages is intimately linked to studying the topology of watersheds.

Issues in Choice of Summit and Key Col

Alteration of the landscape by man and presence of water features can give rise to issues in the choice of location and height of a summit or col. In Britain, extensive discussion has given rise to a protocol that has been adopted by the main sources of prominence data in Britain and Ireland. Other sources of data commonly ignore man-made alterations, but this convention is not universally agreed upon; for example, some authors discount modern structures but allow ancient ones. Another disagreement concerns mountaintop removal, though for high-prominence peaks (and for low-prominence subpeaks with intact summits), the difference in prominence values for the two conventions is typically relatively small.

Interesting Prominence Situations

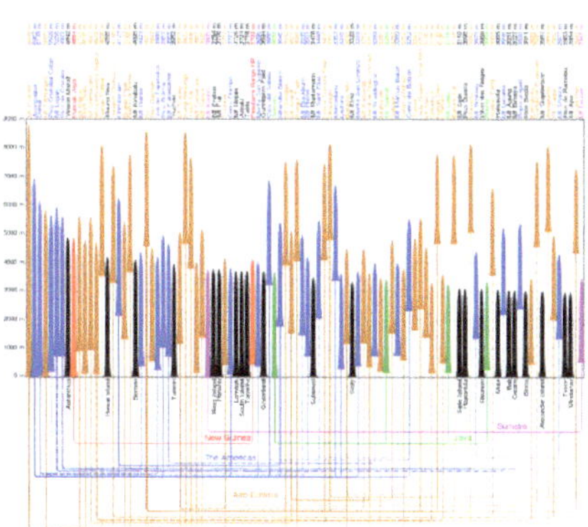

Chart showing relationships between the 100 peaks with highest prominence on Earth.

The key col and parent peak are often close to the subpeak but this is not always the case, especially when the key col is relatively low. It is only with the advent of computer programs and geographical databases that thorough analysis has become possible.

- The key col of Denali in Alaska (6,194 m) is a 56 m col near Lake Nicaragua (unless one accepts the Panama Canal as a key col; this is a matter of contention). Denali's encirclement parent is Aconcagua (6,960 m), in Argentina, and its prominence is 6,138 m. To further illustrate the rising-sea model of prominence, if sea level rose 56 m, North and South America would be separate continents and Denali would be 6138 m above sea level. At a slightly lower level, the continents would still be connected, and the high point of the combined landmass would be Aconcagua, the encirclement parent. Note that, for the purposes of this article, man made structures such as the Panama Canal are not taken into account. If they were, the key col would be along the 26 m Gaillard Cut and Denali would have a prominence of 6,168 m.

While it is natural for Aconcagua to be the parent of Denali, since Denali is a major peak, consider the following situation: Peak A is a small hill on the coast of Alaska, with elevation 100 m and key col 50 m. Then the encirclement parent of Peak A is also Aconcagua, even though there will be many peaks closer to Peak A which are much higher and more prominent than Peak A (for example, Denali). This illustrates the disadvantage in using the encirclement parent.

- Mount Whitney (4421 m) has its key col 1,022 km (635 mi) away in New Mexico at 1347 m on the Continental Divide. Its encirclement parent is Pico de Orizaba (5,636 m), the highest mountain in Mexico. Orizaba's key col is back along the Divide, in British Columbia.

- The key col for Mount Mitchell, the highest peak of the Appalachians, is in Chicago—the low point on the divide between the St. Lawrence and Mississippi River watersheds.

A hill in a low-lying area like the Netherlands will often be a direct child of Mount Everest, with its prominence about the same as its height and its key col placed at or near the foot of the hill, well below, for instance, the 113-meter-high key col of Mont Blanc.

Calculations and Mathematics

When the key col for a peak is close to the peak itself, prominence is easily computed by hand using a topographic map. However, when the key col is far away, or when one wants to calculate the prominence of many peaks at once, a computer is quite useful. Edward Earl has written a program called WinProm which can be used to make such calculations, based on a Digital Elevation Model. The underlying mathematical theory is called "Surface Network Modeling," and is closely related to Morse Theory.

A note about methodology: when using a topographic map to determine prominence, one often has to estimate the height of the key col (and sometimes, the height of the peak as well) based on the contour lines. Assume for simplicity that only the col elevation is uncertain. There are three simple choices. The *pessimistic*, or *clean* prominence, assumes that the col is as high as it can be, i.e., its elevation is that of the higher contour line nearest the saddle. This gives a lower bound on the possible prominence of the peak, ignoring map error; inaccuracies in mapping lead to further uncertainties and a larger error bound. *Optimistic* prominence assumes that the col is as low as possible, yielding an upper bound value for the prominence. *Midrange* or *mean* prominence uses the mean of these two values. *Mean* prominence is sometimes referred to as *rise*. In Britain, where topographic mapping is more detailed than in many countries, it is customary to use a fourth method: *interpolated* prominence. The true prominence is estimated by visualising the three dimensional surface in the neighborhood of the col and interpolating between the enclosing contours.

Which methodology is used depends on the person doing the calculation and on the use to which the prominence is put. For example, if one is making a list of all peaks with at least 2,000 ft (610 m) of prominence, one would might use the optimistic prominence, to include all possible candidates (knowing that some of these could be dropped off the list by further, more accurate, measurements). If one wishes to present the most accurate data for the peaks, mean or interpolated prominence would be appropriate as the other measures give biased estimates.

Wet Prominence and Dry Prominence

There are two varieties of topographic prominence: wet prominence and dry prominence. Wet prominence is the topographic prominence discussed in this article. Wet prominence assumes that the surface of the earth includes all permanent water, snow, and ice features. Thus, the wet prominence of the highest summit of an ocean island or landmass is always equal to the summit's elevation.

Dry prominence, on the other hand, ignores water, snow, and ice features and assumes that the surface of the earth is defined by the solid bottom of those features. The dry prominence of a summit is equal to the wet prominence of that summit unless the summit is the highest point of a landmass or island, or a summit surrounded by snow or ice. If a summit is completely surrounded by a water, snow, or ice feature, the dry prominence of that summit is equal to the wet prominence plus the depth of the highest col.

The dry prominence of Mount Everest is, by convention, equal to its wet prominence (8848 m) plus the depth of the deepest hydrologic feature (the Challenger Deep at 10,911 m), or 19,761 m. The dry prominence of Mauna Kea is equal to its wet prominence (4205 m) plus the depth of its highest col (about 5125 m), or about 9330 m; this is the world's second greatest dry prominence after Mount Everest. The dry prominence of Aconcagua is equal to its wet prominence (6962 m) plus the depth of the highest col of the Bering Strait (about 50 m), or about 7012 m.

Dry prominence is also useful for measuring submerged seamounts. Submerged summits have both a dry topographic prominence and a topographic isolation.

The following mental exercise may illustrate the meaning of topographic prominence. Imagine you are standing at the top of a mountain peak. Let an imaginary sea level slowly rise to the top of your peak. As the imaginary sea level rises, the mountain will shrink into an ever smaller island until the island disappears at your feet. Now let the imaginary sea level slowly drop. The island will reappear with your peak as its highest point. As the sea level continues to drop, the island will grow and merge with adjacent islands. At some point, unless you are atop Mount Everest, your island will touch another island with a higher summit.[b] This point is called your peak's key col, and the elevation rise from its key col to your summit is called its topographic prominence.

Pit-and-mound Topography

Pit and mounds are small, persistent microtopographical features that present themselves after a disturbance event occurs and uproots trees via windthrow. The uprooted tree falls, and a pit forms in the forest floor where the root mass and associated soil matrix used to be. Eventually after a period of time in which the roots decay, the associated soil matrix that was pulled out of the ground with the roots falls back to the ground, creating a corresponding mound.

Formation of a Pit

A pit, as defined above, is formed when pressure is applied to the trunk and crown of the tree that is more powerful than the root and associated soil's ability to hold the tree upright and in place and knocks the tree down. This toppling of the tree can be caused by a multitude of different types of forest disturbance. Wind may blow the tree down; snow may accumulate and put excessive weight on the tree; the tree's roots may have decayed to a point where they are not strong enough to hold the tree upright. Soil conditions also play a role in the ability of the tree to remain upright. Wet soil can loosen the hold a tree's roots have on the soil, and dry soil can help hold the tree in place.

Formation of a Mound

A mound, as defined above, is formed on average from five to ten years after the formation of a pit. The root mass must decay to an extent where the soil matrix that was suspended with it when it toppled over can slough off on to the ground near the corresponding pit. The necessity of the decay of the root mass is what causes this lag time.

In more cold and wet climates, the rate of decay is slower and this time lag may be extended.

Disturbances

Disturbance Scale

Pit and mounds always occur on a fine spatial scale, being the result of only one tree felling. Commonly they are a product of windthrow, but they can also be caused by other factors. Large amounts of snow accumulation on tree boughs or extensive root decay are other possible causes for tree uprooting. Pit and mounds have been analyzed on both on a small scale and larger scale forest systems. It has been observed that they can act as important soil disruptors and microsites for biodiversity and plant establishment.

Microclimatic Conditions

It has been observed that pit and mounds on an individual basis generally have specific climatic and pedological characteristics. Mounds are generally observed to be warmer and drier than the forest floor or the corresponding pit; Similarly, the pit is generally found to be colder and have a higher soil moisture content than either the forest floor or the mound. One notable instance where there is an exception to the general observation is when a snow layer has fallen. The pit becomes insulated by the snow layer and then in some instances is observed to be warmer than the mound.

Mounds receive the highest amount of photosynthetically active radiation (PAR) and pits the lowest amount. Pits also have much higher amounts of leaf litter than mounds, who have high tendencies to erode.

Nutrient Content and Plant Establishment

Studies on pit and mounds generally have at least five sampling sites per each pit and mound sampled. These areas, generally, are the top of the mound, bottom of the pit, side of the mound, edge of the pit, and an undisturbed forest floor area. Studies have observed that both pits and mounds generally have lower carbon and nitrogen soil content than the undisturbed forest floor, although they have carbon to nitrogen ratios that are not significantly different from the ratios observed on the forest floor.

Mounds are observed to be generally more nutrient poor and fewer plants establish on them in comparison to pits for several reasons. Mounds tend to erode and thus are more unstable than the forest floor or the pit. The surface of the mound also contains very little organic matter because it is mainly soil that was uprooted from the mineral horizons of the soil layer. This coupled with the observation that seed deposition rates are lower for mounds than pits, makes plant establishment on mounds unlikely and problematic for the plant.

Conversely, many studies have found that species richness and overall number of established plants is found to be statistically higher on mounds than in pits. It was experimentally determined that the leaf litter accumulation that occurs within a pit has a large effect on the lack of species diversity or establishment that occurs there. In fact, when scientists removed the leaf litter and monitored the species establishment within the pit, it was found that the diversity and number began to resemble that that occurred on mounds. Multiple studies have found that on the forest floor a high or dense amount of leaf litter or dead organic matter corresponds to a low amount of species diversity as well.

Biodiversity

Macrofauna

Little information is available on how macrofauna are affected by the presence of pit and mounds in a forested landscape. One would believe they would benefit from the presence of coarse woody debris as habitat and new flora that establish in the pit and mounds, but there is a lack of published evidence.

One macrofauna species that has received attention is the earthworm population in a lowland temperate rainforest of Belgium. It was observed that both earthworm speciation and biomass differed based on location within the pit and mound microsites examined. Undisturbed sites near the pit and mounds had the highest amount of species diversity and biomass compared to either a pit or a mound. The undisturbed sites contained species of earthworm that were not present in either the pit or mound sample site. An explanation towards why there were less earthworms present in the pit or mound was that within the mound there was more dry, mineralized soil present and less organic matter due the uprooting of the tree. The pits examined were found to generally only contain one species of earthworm that was known to inhabit very moist soils. It was concluded that the pit microhabitat was too humid and water containing for most other species of earthworm.

Knowledge Gaps

What aspects of pit and mounds can be generalized and what effects they exert of the dynamism of the forest can be generalized is a currently evolving narrative. As more scientific literature crops up from around the world, it becomes more apparent what aspects of the pit and mounds occur in most any biome they are found in. As it is, there is not enough scientific data to allow much extrapolation from forest to forest in regard to pit and mound knowledge.

Location of Experimentation and Data Collection

It should be noted that virtually all of the scientific knowledge about pit and mounds and their impact on forest heterogeneity and other dynamics comes from the Northern Hemisphere, particularly the Relic forests of Europe and the Temperate rainforests of

Eastern Canada and Southeast Alaska. It can be argued that because pit and mounds occur on such a small scale, they may impact forest in the southern hemisphere in a different way. Until information is collected, it will be unknown.

Herbivory

Studies of pit and mound on a large time scale may be affected by the presence of herbivores in the area. A herbivore may preferentially browse certain seedlings that grow on either a pit or a mound and will mask any other variables that might be contributing to why those seedlings are establishing in a pit or mound by skewing the final counts of the growing plants.

Microbiological Communities

One would assume that on a microbial level a disturbance that results in a pit and mound event would lead to a greater heterogeneity on the microbial level as multiple soil horizons are uplifted and mixed. However, little scientific data exists on the microbial population and its changes due to a pit and mound creation.

Impact on Silviculture

There is no scientific data on the effects, if any, of either the synthetic or natural creation of pit and mounds in a managed forest and the effects it has on the growth or health of the forest.

Phytogeomorphology

Phytogeomorphology is the study of how terrain features affect plant growth. It was the subject of a treatise by Howard and Mitchell in 1985, who were considering the growth and varietal temporal and spatial variability found in forests, but recognized that their work also had application to farming, and the relatively new science (at that time) of precision agriculture. The premise of Howard and Mitchell is that landforms, or features of the land's 3D topography significantly affect how and where plants (or trees in their case) grow. Since that time, our ability to map and classify landform shapes and features has increased greatly. The advent of GPS has made it possible to map almost any variable one might wish to measure. Thus, a very increased awareness of the spatial variability of the environment that plants gown in has arisen. The development of technology like airborne LiDAR has enabled the detailed measurement of landform features to better than sub-meter, and when combined with RTK-GPS (accuracies to 1mm) enables the creation of very accurate maps of where these features are. Comparison of these landform maps with mapping of variables related to crop or plant growth show a strong correlation.

Phytogeomorphology and Precision Agriculture

While phytogeomorphology is a broad term that applies to the relationship between plants and terrain attributes in general, it is also very ap-plicable to the aspect of precision agriculture that studies crop growth temporal and spatial variability within farm fields. There is already a volume of work, although they don't use the term phytogeomorphology specifically, that considers farm field terrain attributes as affecting crop yield and growth, Moore et *al.* (1991) provide an early overview of the application of terrain features to precision agriculture, but one of the earliest references to this phenomenon in farming is that of Whittaker in 1967. More recent work includes a six-year study of temporal and spatial yield stability over 11 years (Kaspar et al., (2003), and references therein), and a detailed study of the same on a small patch farm in Portugal (and references therein). This variability can be exploited to produce higher yields and reduce the environmental impact of farming - consequently returning a higher profit to the farmer in terms of higher overall yields and lesser amounts of inputs. The new science of Sustainable Intensification of Agriculture which is addressing the need for higher yields from existing fields can be fulfilled by some of the practical applications of phytogeomorphology applied to precision agriculture.

Work in this area has been happening for some years, Marquas de Silva et al., (2008), and especially Moore et al., (1991)), but it is slow and sometimes tedious work that necessarily involves multiple years of data, very specialized software tools, and long compute times to produce the resulting maps.

Phytogeomorphologically Defined Management Zones

Typically, the objective of precision agriculture is to divide the farm field into distinct management zones based on yield performance at each point in the field. 'Variable rate technology' is a relatively new term in farming technology that refers to spreaders, seeders, sprayers, etc. that are able to adjust their rates of flow on the fly. The idea is to create a 'recipe map' for variable rate farm machinery to deliver the exact quantity of amendments required at that location (within that zone of the field). The literature is divided on how to properly define management zones, and farmers always seem to have their own original ideas (Personal communication, Willemse (2011)).

In the geomorphological approach to defining management zones it is found that topography aids in at least partially defining how much yield comes from which part of the field. This is true in fields where there are permanently limiting characteristics to parts of the field, but not true in fields where the growth potential is the same all over the field (Blackmore et al., (2003)). It can be shown that an index map of yield (shows areas of consistent over-performance of yield and areas of consistent under-performance) correlates well with a landform classification map (personal communication, Aspinall (2011)). Landforms can be classified a number of ways, but the simplest to use software tool is LandMapR (MacMillan (2003)). An early version of the LandMapR software is available through the Opengeomorphometry project hosted under the Google Code project.

References

- Chan, Alan Kam-leung and Gregory K. Clancey, Hui-Chieh Loy (2002). Historical Perspectives on East Asian Science, Technology and Medicine. Singapore: Singapore University Press. p. 15. ISBN 9971-69-259-7.
- Leeder, M., 1999, Sedimentology and Sedimentary Basins, From Turbulence to Tectonics, Blackwell Science, 592 p. ISBN 0-632-04976-6.
- Bennett, M.R. & Glasser, N.F., 1996, Glacial Geology: Ice Sheets and Landforms, John Wiley & Sons Ltd, 364 p. ISBN 0-471-96345-3.
- Dawson, Alan (1997). The Hewitts and Marilyns of England. Glasgow: TACit Press. There are several related booklets covering Britain and Ireland. ISBN 0-9522680-7-8.

3

Landform and its Types

Landforms are features of the solid surface of the Earth. The arrangement of landforms is referred to as topography. Typically, landforms include hills, mountains, canyons, valleys and mid-ocean ridges. The topics discussed in the chapter are of great importance to broaden the existing knowledge on landforms.

Landform

A landform is a natural feature of the solid surface of the Earth or other planetary body. Landforms together make up a given terrain, and their arrangement in the landscape is known as topography. Typical landforms include hills, mountains, plateaus, canyons, valleys, as well as shoreline features such as bays, peninsulas, and seas, including submerged features such as mid-ocean ridges, volcanoes, and the great ocean basins.

This conical hill in Argentina called Cono de Arita constitutes a landform.

Physical Characteristics

Landforms are categorized by characteristic physical attributes such as elevation, slope, orientation, stratification, rock exposure, and soil type. Gross physical features or landforms include intuitive elements such as berms, mounds, hills, ridges, cliffs, valleys, rivers, peninsulas and numerous other structural and size-scaled (i.e. ponds vs. lakes, hills vs. mountains) elements including various kinds of inland and oceanic waterbodies and sub-surface features.

This panorama in Great Smoky Mountains National Park has the readily identifiable **physical features** of a rolling plain, actually part of a broad valley, distant foothills, and a backdrop of the old, much weathered Appalachian mountain range

Hierarchy of Classes

Karst towers landforms along Lijiang River, Guilin, China

Oceans and continents exemplify the highest-order landforms. Landform elements are parts of a high-order landforms that can be further identified and systematically given a cohesive definition such as hill-tops, shoulders, saddles, foreslopes and backslopes.

Some generic landform elements including: pits, peaks, channels, ridges, passes, pools and plains.

Terrain (or *relief*) is the third or vertical dimension of *land surface*. Topography is the study of terrain, although the word is often used as a synonym for relief itself. When relief is described underwater, the term bathymetry is used. In cartography, many different techniques are used to describe relief, including contour lines and TIN (Triangulated irregular network).

Elementary landforms (segments, facets, relief units) are the smallest homogeneous divisions of the land surface, at the given scale/resolution. These are areas with relatively homogeneous morphometric properties, bounded by lines of discontinuity. A plateau or a hill can be observed at various scales ranging from few hundred meters to hundreds of kilometers. Hence, the spatial distribution of landforms is often scale-dependent as is the case for soils and geological strata.

A number of factors, ranging from plate tectonics to erosion and deposition, can generate and affect landforms. Biological factors can also influence landforms— for example, note the role of vegetation in the development of dune systems and salt marshes, and the work of corals and algae in the formation of coral reefs.

Landforms do not include man-made features, such as canals, ports and many harbors; and geographic features, such as deserts, forests, and grasslands. Many of the terms are not restricted to refer to features of the planet Earth, and can be used to describe surface features of other planets and similar objects in the Universe. Examples are mountains, hills, polar caps, and valleys, which are found on all of the terrestrial planets.

Recent Developments

Landforms may be extracted from a digital elevation model using some automated techniques where the data has been gathered by modern satellites and stereoscopic aerial surveillance cameras. Until recently, compiling the data found in such data sets required time consuming and expensive techniques involving many man-hours. The most detailed DEMs available are measured directly using LIDAR techniques.

Hill

A hill is a landform that extends above the surrounding terrain. It often has a distinct summit, although in areas with scarp/dip topography a hill may refer to a particular section of flat terrain without a massive summit (e.g. Box Hill, Surrey).

Terminology

A khulan (Mongolian wild ass) on a hill in the Gobi of Mongolia at sunset

Chocolate Hills of the Philippines

Hills in Tuscany, Italy

The distinction between a hill and a mountain is unclear and largely subjective, but a hill is universally considered to be less tall and less steep than a mountain. In the United Kingdom, geographers historically regarded mountains as hills greater than 1,000 feet (300 m) above sea level, which formed the basis of the plot of the 1995 film *The Englishman Who Went Up a Hill But Came Down a Mountain*. In contrast, hillwalkers have tended to regard mountains as peaks 2,000 feet (610 m) above sea level: the *Oxford English Dictionary* also suggests a limit of 2,000 feet (610 m) and Whittow states "Some authorities regard eminences above 600 m (2,000 ft) as mountains, those below being referred to as hills." The Great Soviet Encyclopedia defines hill as an upland with a relative height up to 200 m (660 ft).

Today, a mountain is usually defined in the UK and Ireland as any summit at least 2,000 feet (or 610 meters) high, while the official UK government's definition of a mountain is a summit of 600 meters or higher. Some definitions include a topographical prominence requirement, typically 100 feet (30 m) or 500 feet (152 m). In practice, mountains in Scotland are frequently referred to as "hills" no matter what their height, as reflected in names such as the *Cuillin Hills* and the *Torridon Hills*. In Wales, the distinction is more a term of land use and appearance and has nothing to do with height. For a while, the U.S. defined a mountain as being 1,000 feet (304.8 m) or more tall. Any similar landform lower than this height was considered a hill. The United States Geological Survey (USGS), however, has concluded that these terms do not in fact have technical definitions in the U.S.

A hillock is a small hill. Other words include knoll and (in Scotland, Northern Ireland and northern England) its variant, knowe. Artificial hills may be referred to by a variety of technical names, including mound and tumulus.

Hills may form through geomorphic phenomena: faulting, erosion of larger landforms, such as mountains and movement and deposition of sediment by glaciers (e.g. moraines and drumlins or by erosion exposing solid rock which then weathers down into a hill.) The rounded peaks of hills results from the diffusive movement of soil and regolith covering the hill, a process known as downhill creep.

Hills of the Judean Desert

Various names used to describe types of hill, based on appearance and method of formation. Many such names originated in one geographical region to describe a type of hill formation peculiar to that region, though the names are often adopted by geologists and used in a wider geographical context. These include:

- Drumlin – an elongated whale-shaped hill formed by glacial action.
- Butte – an isolated hill with steep sides and a small flat top, formed by weathering.
- *Kuppe* – a rounded hill or low mountain, typical of central Europe
- Tor – a rock formation found on a hilltop; also used to refer to the hill, especially in South West England.
- Puy – used especially in the Auvergne, France, to describe a conical volcanic hill.
- Pingo – a mound of earth-covered ice found in the Arctic and Antarctica.

Historical Significance

Clouds over hills in Steptoe, Washington

Many settlements were originally built on hills, either to avoid or curb floods, particularly if they were near a large body of water, or for defense, since they offer a good view of the surrounding land and require would-be attackers to fight uphill. For example, Ancient Rome was built on seven hills, protecting it from invaders.

In northern Europe, many ancient monuments are sited in heaps. Some of these are defensive structures (such as the hill-forts of the Iron Age), but others appear to have hardly any significance. In Britain, many churches at the tops of hills are thought to have been built on the sites of earlier pagan holy places. The National Cathedral in Washington, DC has followed this tradition and was built on the highest hill in that city.

Military Significance

The Battle of Bunker Hill

Hills provide a major advantage to an army, giving them an elevated firing position and forcing an opposing army to charge uphill to attack them. They may also conceal forces behind them, allowing a force to lay in wait on the crest of a hill, using that crest for cover, and firing on unsuspecting attackers as they broach the hilltop. As a result, conventional military strategies often demand possession of high ground.

Hills have become sites for many noted battles, such as the first recorded military conflict in Scotland known as the battle of Mons Graupius. Modern conflicts include the Battle of Bunker Hill (which was actually fought on Breed's Hill) in the American War for Independence and Cemetery Hill and Culp's Hill in the Battle of Gettysburg, the turning point of the American Civil War.

Role in shaping of a country for a combined United States of America: In 1864, 160,000 Union and Confederate soldiers fought at Pigeon Hill and Cheatham Hill Battlefield on the land (with trenches and cannons) that now forms the National Battlefield Park at Kennesaw, in Georgia; when Confederate General Joseph E. Johnston and his troops temporarily halted Union General William T. Sherman's advance toward Atlanta from North. "This is hallowed ground. For two terrible weeks in the summer of 1864, thousands of Americans on both sides fought and died here. The fierce struggle they waged here not only helped decide the outcome of American Civil War, it also shaped the fate of a nation, and destiny of generations yet unborn."

The Battle of San Juan Hill in the Spanish–American War won Americans control of Santiago. The Battle of Alesia was also fought from a hilltop fort. Fighting on Mamayev Kurgan during the Battle of Stalingrad and the Umurbrogol Pocket in the Battle of Peleliu were also examples of bloody fighting for high ground. Another recent example is the Kargil War between India and Pakistan. The Great Wall of China is also an example of an advantage provider. It is built on mountain tops, and was meant to defend against invaders from the north, and among others, Mongolians.

Sports and Games

Hillwalking is a British English term for a form of hiking which involves the ascent of hills. The activity is usually distinguished from mountaineering as it does not involve ropes or technically difficult rock climbing, although the terms mountain and hill are often used interchangeably in Britain. Hillwalking is popular in mountainous areas such as the English Peak District or the Scottish Highlands. Many hills are categorized according to relative height or other criteria and feature on lists named after mountaineers, such as Munros (Scotland) or Wainwrights (England). Specific locating activities such as "peak bagging" (or "Munro bagging") involve climbing hills on these lists with the aim of completing (or "compleating") the list.

Hillwalkers on Beinn Dearg, Scotland

An example of a golf course in England that has hills

In golf, the terrain on golf courses is often made more rugged and hilly to make the holes harder to play. For example, the hole may be located at the top of a hill, and the course is designed specifically to make it almost impossible to allow the golf ball to rest near the top; it would roll down, and the player would have to try again.

Cheese rolling is an annual event in the West Country of England which involves rolling a wheel of cheese down a hill. Contestants stand at the top and chase the wheel of cheese to the bottom. The winner, the one who catches the cheese, gets to keep the wheel of cheese as a prize.

Largest man-made

- Sophienhöhe, Germany (200 m)
- Kvarntorpshogen, Sweden (100 m)
- Expedition Everest (61 m (200 ft))
- Blackstrap Ski Hill (45 m (148 ft))
- Silbury Hill (40 m (130 ft))
- Grizzly Peak (Disneyland Resort) (34 m (112 ft))
- Mount Manisty (100 ft)
- Monks Mound (30 m (98 ft))
- Mount Gushmore (27 m (89 ft))

Mountain

A mountain is a large landform that stretches above the surrounding land in a limited area, usually in the form of a peak. A mountain is generally steeper than a hill. Mountains are formed through tectonic forces or volcanism. These forces can locally raise the surface of the earth. Mountains erode slowly through the action of rivers, weather conditions, and glaciers. A few mountains are isolated summits, but most occur in huge mountain ranges.

Mount Everest, the highest peak on Earth

Chimborazo, Ecuador. The point on Earth's surface farthest from its center.

Aiguille du Dru in the French Alps

High elevations on mountains produce colder climates than at sea level. These colder climates strongly affect the ecosystems of mountains: different elevations have different plants and animals. Because of the less hospitable terrain and climate, mountains tend to be used less for agriculture and more for resource extraction and recreation, such as mountain climbing.

The highest mountain on Earth is Mount Everest in the Himalayas of Asia, whose summit is 8,850 m (29,035 ft) above mean sea level. The highest known mountain on any planet in the Solar System is Olympus Mons on Mars at 21,171 m (69,459 ft).

Definition

There is no universally accepted definition of a mountain. Elevation, volume, relief, steepness, spacing and continuity have been used as criteria for defining a mountain. In the Oxford English Dictionary a mountain is defined as "a natural elevation of the earth surface rising more or less abruptly from the surrounding level and attaining an altitude which, relatively to the adjacent elevation, is impressive or notable."

The Matterhorn, Swiss Alps

Peaks of Mount Kenya

Whether a landform is called a mountain may depend on local usage. The highest point in San Francisco, California, is called Mount Davidson, notwithstanding its height of 300 m (980 ft), which makes it twenty feet short of the minimum for a mountain by American designations. Similarly, Mount Scott outside Lawton, Oklahoma is only 251 m (823 ft) from its base to its highest point. Whittow's *Dictionary of Physical Geography* states *"Some authorities regard eminences above 600 m (2,000 ft) as mountains, those below being referred to as hills."*

In the United Kingdom and the Irish Republic, a mountain is usually defined as any summit at least 2,000 feet (or 610 metres) high, whilst the official United Kingdom government's definition of a mountain, for the purposes of access, is a summit of 600 metres or higher. In addition, some definitions also include a topographical prominence requirement, typically 100 or 500 feet (30 or 152 m). For a while, the US defined a mountain as being 1,000 feet (300 m) or taller. Any similar landform lower than this height was considered a hill. However, today, the United States Geological Survey (USGS) concludes that these terms do not have technical definitions in the US.

The UN Environmental Programme's definition of "mountainous environment" includes any of the following:

- Elevation of at least 2,500 m (8,200 ft);
- Elevation of at least 1,500 m (4,900 ft), with a slope greater than 2 degrees;
- Elevation of at least 1,000 m (3,300 ft), with a slope greater than 5 degrees;
- Elevation of at least 300 m (980 ft), with a 300 m (980 ft) elevation range within 7 km (4.3 mi).

Using these definitions, mountains cover 33% of Eurasia, 19% of South America, 24% of North America, and 14% of Africa. As a whole, 24% of the Earth's land mass is mountainous.

Geology

There are three main types of mountains: volcanic, fold, and block. All three types are formed from plate tectonics: when portions of the Earth's crust move, crumple, and dive. Compressional forces, isostatic uplift and intrusion of igneous matter forces surface rock upward, creating a landform higher than the surrounding features. The height of the feature makes it either a hill or, if higher and steeper, a mountain. Major mountains tend to occur in long linear arcs, indicating tectonic plate boundaries and activity.

Jeff Davis Peak seen from the glacier-carved summit of Wheeler Peak, Nevada

Volcanoes

Volcanoes are formed when a plate is pushed below another plate, or at a mid-ocean ridge or hotspot. At a depth of around 100 km, melting occurs in rock above the slab (due to the addition of water), and forms magma that reaches the surface. When the magma reaches the surface, it often builds a volcanic mountain, such as a shield volcano or a stratovolcano. Examples of volcanoes include Mount Fuji in Japan and Mount Pinatubo in the Philippines. The magma does not have to reach the surface in order to create a mountain: magma that solidifies below ground can still form dome mountains, such as Navajo Mountain in the United States.

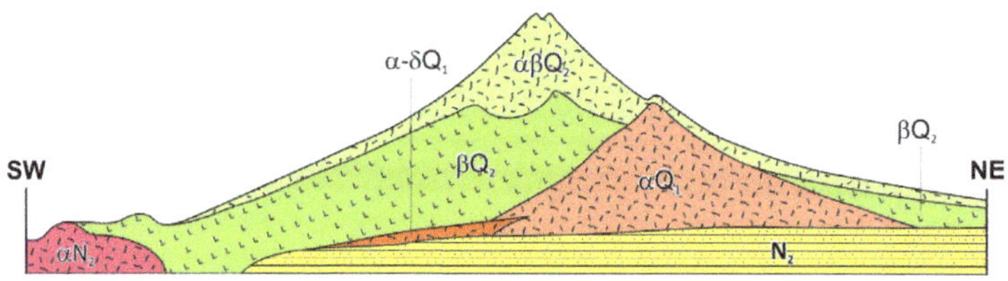

Geological cross-section of Fuji volcano

Fold Mountains

Fold mountains occur when two plates collide: shortening occurs along thrust faults and the crust is overthickened. Since the less dense continental crust "floats" on the denser mantle rocks beneath, the weight of any crustal material forced upward to form hills, plateaus or mountains must be balanced by the buoyancy force of a much greater volume forced downward into the mantle. Thus the continental crust is normally much thicker under mountains, compared to lower lying areas. Rock can fold either symmetrically or asymmetrically. The upfolds are anticlines and the downfolds are synclines: in asymmetric folding there may also be recumbent and overturned folds. The Jura Mountains are an example of fold mountains.

Block Mountains

The Catskills in Upstate New York represent an eroded plateau.

Block mountains are caused by faults in the crust: a seam where rocks can move past each other. When rocks on one side of a fault rise relative to the other, it can form a mountain. The uplifted blocks are block mountains or horsts. The intervening dropped blocks are termed graben: these can be small or form extensive rift valley systems. This form of landscape can be seen in East Africa, the Vosges, the Basin and Range Province of Western North America and the Rhine valley. These areas often occur when the regional stress is extensional and the crust is thinned.

Erosion

During and following uplift, mountains are subjected to the agents of erosion (water, wind, ice, and gravity) which gradually wear the uplifted area down. Erosion causes the surface of mountains to be younger than the rocks that form the mountains themselves. Glacial processes produce characteristic landforms, such as pyramidal peaks, knife-edge arêtes, and bowl-shaped cirques that can contain lakes. Plateau mountains, such as the Catskills, are formed from the erosion of an uplifted plateau.

Climate

Climate on mountains become colder at high elevations, due an interaction between radiation and convection. Sunlight in the visible spectrum hits the ground and heats it. The ground then heats the air at the surface. If radiation were the only way to transfer heat from the ground to space, the greenhouse effect of gases in the atmosphere would keep the ground at roughly 333 K (60 °C; 140 °F), and the temperature would decay exponentially with height.

A mountain in Carbon County, Utah

Valley of the Ten Peaks, Canadian Rockies

However, when air is hot, it tends to expand, which lowers its density. Thus, hot air tends to rise and transfer heat upward. This is the process of convection. Convection comes to equilibrium when a parcel at air at a given altitude has the same density as its surroundings. Air is a poor conductor of heat, so a parcel of air will rise and fall without exchanging heat. This is known as an adiabatic process, which has a characteristic pressure-temperature curve. As the pressure gets lower, the temperature decreases. The rate of decrease of temperature with elevation is known as the adiabatic lapse rate, which is approximately 9.8 °C per kilometer (or 5.4 °F per 1000 feet) of altitude.

Note that the presence of water in the atmosphere complicates the process of convection. Water vapor contains latent heat of vaporization. As air rises and cools, it eventually becomes saturated and cannot hold its quantity of water vapor. The water vapor condenses (forming clouds), and releases heat, which changes the lapse rate from the dry adiabatic lapse rate to the moist adiabatic lapse rate (5.5 °C per kilometer or 3 °F per 1000 feet) The actual lapse rate can vary by altitude and by location.

Therefore, moving up 100 meters on a mountain is roughly equivalent to moving 80 kilometers (45 miles or 0.75° of latitude) towards the nearest pole. This relationship is only approximate, however, since local factors such as proximity to oceans (such as the Arctic Ocean) can drastically modify the climate. As the altitude increases, the main form of precipitation becomes snow and the winds increase.

The effect of the climate on the ecology at an elevation can be largely captured through a combination of amount of precipitation, and the biotemperature, as described by Leslie Holdridge in 1947. Biotemperature is the mean temperature; all temperatures below 0 °C (32 °F) are considered to be 0 °C. When the temperature is below 0 °C, plants are dormant, so the exact temperature is unimportant. The peaks of mountains with permanent snow can have a biotemperature below 1.5 °C (34.7 °F).

Ecology

The colder climate on mountains affects the plants and animals residing on mountains. A particular set of plants and animals tend to be adapted to a relatively narrow range of climate. Thus, ecosystems tend to lie along elevation bands of roughly constant climate. This is called altitudinal zonation. In regions with dry climates, the tendency of mountains to have higher precipitation as well as lower temperatures also provides for varying conditions, which enhances zonation.

An alpine mire in the Swiss Alps

Some plants and animals found in altitudinal zones tend to become isolated since the conditions above and below a particular zone will be inhospitable and thus constrain their movements or dispersal. These isolated ecological systems are known as sky islands.

Altitudinal zones tend to follow a typical pattern. At the highest elevations, trees cannot grow, and whatever life may be present will be of the alpine type, resembling tundra. Just below the tree line, one may find subalpine forests of needleleaf trees, which can withstand cold, dry conditions. Below that, montane forests grow. In the temperate portions of the earth, those forests tend to be needleleaf trees, while in the tropics, they can be broadleaf trees growing in a rain forest.

In Society

Mountains are generally less preferable for human habitation than lowlands, because of harsh weather and little level ground suitable for agriculture. While 7% of the land area of Earth is above 2,500 metres (8,200 ft), only 140 million people live above that altitude and only 20-30 million people above 3,000 metres (9,800 ft) elevation. The decreasing atmospheric pressure with increasing elevation means that less oxygen is available for breathing, and there is less protection against solar radiation (UV). Due to decreasing oxygen, the highest known permanent habitation in the world is at 5,100 metres (16,700 ft), while the highest known permanently tolerable altitude is at 5,950 metres (19,520 ft). Above 8,000 metres (26,000 ft) elevation, there is not enough oxygen to support human life. This is known as the "death zone". The summits of Mount Everest and K2 are in the death zone.

Mountain climbers ascending Mount Rainier

The summit of Ben Nevis, the British Isles' highest, has a memorial

About half of mountain dwellers live in the Andes, Central Asia, and Africa. Traditional mountain societies rely on agriculture, with higher risk of crop failure than at lower elevations. Minerals often occur in mountains, with mining being an important component of the economics of some montane societies. More recently, tourism supports mountain communities, with some intensive development around attractions such as national parks or ski resorts. About 80% of mountain people live below the poverty line.

Most of the world's rivers are fed from mountain sources, with snow acting as a storage mechanism for downstream users. More than half of humanity depends on mountains for water.

Mountaineering, mountain climbing, or alpinism is the sport, hobby or profession of hiking, skiing, and climbing mountains. While mountaineering began as attempts to reach the highest point of unclimbed big mountains it has branched into specializations that address different aspects of the mountain and consists of three areas: rock-craft, snow-craft and skiing, depending on whether the route chosen is over rock, snow or ice. All require experience, athletic ability, and technical knowledge to maintain safety.

Superlatives

The Zugspitze, the highest mountain in Germany

Maat Mons of Venus (22.5x exaggeration)

Heights of mountains are typically measured above sea level. Using this metric, Mount Everest is the highest mountain on Earth, at 8,848 metres (29,029 ft). There are at least 100 mountains with heights of over 7,200 metres (23,622 ft) above sea level, all of which are located in central and southern Asia. The highest mountains above sea level are generally not the highest above the surrounding terrain. There is no precise definition of surrounding base, but Denali, Mount Kilimanjaro and Nanga Parbat are possible candidates for the tallest mountain on land by this measure. The bases of mountain islands are below sea level, and given this consideration Mauna Kea (4,207 m (13,802 ft) above sea level) is the world's tallest mountain and volcano, rising about 10,203 m (33,474 ft) from the Pacific Ocean floor.

The highest mountains are not generally the most voluminous. Mauna Loa (4,169 m or 13,678 ft) is the largest mountain on Earth in terms of base area (about 2,000 sq mi or 5,200 km^2) and volume (about 18,000 cu mi or 75,000 km^3). Mount Kilimanjaro is the largest non-shield volcano in terms of both base area (245 sq mi or 635 km²) and volume (1,150 cu mi or 4,793 km³). Mount Logan is the largest non-volcanic mountain in base area (120 sq mi or 311 km²).

The highest mountains above sea level are also not those with peaks farthest from the centre of the Earth, because the figure of the Earth is not spherical. Sea level closer to the equator is several miles farther from the centre of the Earth. The summit of Chimborazo, Ecuador's tallest mountain, is usually considered to be the farthest point from the Earth's centre, although the southern summit of Peru's tallest mountain, Huascarán, is another contender. Both have elevations above sea level more than 2 kilometres (6,600 ft) less than that of Everest.

Canyon

A canyon (Spanish: cañón) or gorge is a deep ravine between pairs of escarpments or cliffs and is the most often carved landscape by the erosive activity of a river over geologic timescales. Rivers have a natural tendency to cut through underlying surfaces so will eventually wear away rock layers to lessen their own pitch slowing their waters; given enough time, their bottoms will gradually reach a baseline elevation—which is the same elevation as the body of water it will eventually drain into. This action, when the river source and mouth are at much different base elevations will form a canyon, particularly through regions where softer rock layers are intermingled with harder layers more resistant to weathering.

The Grand Canyon, Arizona, at the confluence of the Colorado and Little Colorado rivers

A canyon may also refer to a rift between two mountain peaks, such as those in ranges including the Rocky Mountains, the Alps, the Himalayas or the Andes. Usually a river or stream and erosion carve out such splits between mountains. Examples of mountain-type canyons are Provo Canyon in Utah or Yosemite National Park in California's Sierra Nevada. Canyons within mountains, or gorges that have an opening on only one side are called *box canyons*. Slot canyons are very narrow canyons, often with smooth walls.

Boğazpınar Canyon, Mersin Province, Turkey

Itaimbezinho Canyon, at Aparados da Serra National Park, Brazil

One of three gorges on Yangtze river, China

Steep-sided valleys in the seabed of the continental slope underwater are referred to as submarine canyons. Unlike canyons on land, submarine canyons are thought to be formed by turbidity currents and underwater landslides.

Etymology

The word *canyon* is Spanish in origin (*cañón*, pronounced: [ka□□on], with the same meaning. The word *canyon* is generally used in North America while the words *gorge* and *ravine* are used in Europe and Oceania, though *gorge* and *ravine* are also used in some parts of North America. In the United States, place names generally use "canyon" in the southwest and "gorge" in the northeast, with the rest of the country graduating between these two according to geography. In Canada, "gorges" are usually narrow, and "ravines" more open and often wooded. The military-derived word *defile* is occasionally used in the United Kingdom.

Formation

Most canyons were formed by a process of long-time erosion from a plateau or table-land level. The cliffs form because harder rock strata that are resistant to erosion and weathering remain exposed on the valley walls.

Canyons are much more common in arid than in wet areas because physical weathering has a more localized effect in arid zones. The wind and water from the river combine to erode and cut away less resistant materials such as shales. The freezing and expansion of water also serves to help form canyons. Water seeps into cracks between the rocks and freezes, pushing the rocks apart and eventually causing large chunks to break off the canyon walls, in a process known as frost wedging. Canyon walls are often formed of resistant sandstones or granite.

Sometimes large rivers run through canyons as the result of gradual geological uplift. These are called entrenched rivers, because they are unable to easily alter their course. In the United States, the Colorado River in the Southwest and the Snake River in the Northwest are two examples of tectonic uplift.

Canyons often form in areas of limestone rock. As limestone is soluble to a certain extent, cave systems form in the rock. When these collapse, a canyon is left, as in the Mendip Hills in Somerset and Yorkshire Dales in Yorkshire, England.

Box Canyon

A box canyon is a small ravine or canyon with steep walls on three sides, allowing access and egress only through the mouth of the canyon. Box canyons were frequently used in the American West as convenient corrals, with their entrances fenced. They were also used as kill sites for wild game, which could be driven into the confined space and killed.

Largest Canyons

The definition of "largest canyon" is imprecise, because a canyon can be large by its depth, its length, or the total area of the canyon system. Also, the inaccessibility of the major canyons in the Himalaya contributes to their not being regarded as candidates for the biggest canyon. The definition of "deepest canyon" is similarly imprecise, especially if one includes mountain canyons as well as canyons cut through relatively flat plateaus (which have a somewhat well-defined rim elevation).

The deepest canyon in the world, the Indus Gorge with Nanga Parbat, the world's 9th highest mountain, rising to the south.

The deepest canyon in the world is at the confluence of the Gilgit and Indus rivers, where the Karakoram Highway was constructed; the river flows at an elevation of about 1,000 m (3,300 ft) above sea level. To the south by 15 km, rising in one continuous sweep above the Indus, the world's largest such rise is the summit of the world's ninth-highest mountain, Nanga Parbat, 8,126 m (26,660 ft) high. Rising to the north, also visible from the bridge over the Gilgit, is the westernmost of the great Himalayan peaks, Rakaposhi, 7,788 m (25,550 ft) high. That makes the Indus gorge 7,120 m (23,360 ft) deep from river to peak at that point, more than 600 m (2,000 ft) greater than the depth of the Kali Gandaki Gorge and 1,500 m (5,000 ft) deeper than the Yarlung Tsangpo Gorge. These are the three deepest canyons in the world. Only the largest mountain ranges of Asia have the scale for such canyons to have formed. The Andes of South America are high mountains, but not on the same scale as the Himalayas, Karakoram, or Hindu Kush, which all come together at the confluence of the Gilgit and the Indus in Pakistani Kashmir.

The Yarlung Tsangpo Grand Canyon (or Tsangpo Canyon), along the Yarlung Tsangpo River in Tibet, is regarded by some as the deepest canyon in the world at 5,500 m (18,000 ft). It is slightly longer than the Grand Canyon in the United States. Others consider the Kali Gandaki Gorge in midwest Nepal to be the deepest canyon, with a 6400 m (21,000 ft) difference between the level of the river and the peaks surrounding it.

Vying for deepest canyon in the Americas are the Cotahuasi Canyon and Colca Canyon, in southern Peru. Both have been measured at over 3500 m (12,000 ft) deep.

View of the Grand Canyon from the Space Shuttle Challenger

The Grand Canyon of northern Arizona in the United States, with an average depth of 1,600 m (one mile) and a volume of 4.17 trillion cubic metres, is one of the world's largest canyons. It was among the 28 finalists of the New7Wonders of Nature worldwide poll. (Some referred to it as one of the 7 natural wonders of the world.) Copper Canyon in Chihuahua, Mexico is deeper and longer than the Grand Canyon.

The largest canyon in Africa is the Fish River Canyon in Namibia.

In August 2013, the discovery of Greenland's Grand Canyon was reported. Located under an ice sheet, based on the analysis of date from Operation IceBridge, it is believed to be the longest canyon in the world.

The Capertee Valley in Australia is commonly reported as being the second largest (in terms of width) of any canyon in the world.

Panoramic view of the Capertee Valley in Australia, the second largest (in terms of width) of any canyon in the world

Culturce

Some canyons have notable cultural significance. Evidence of early humanoids has been discovered in Africa's Olduvai Gorge. In the southwestern United States, canyons are important archeologically because of the many cliff-dwellings built in such areas, largely by the ancient Pueblo people who were their first inhabitants.

List of notable Canyons and Gorges

Africa

- Blyde River Canyon, Mpumalanga, South Africa
- Fish River Canyon, Namibia
- Kloof, KwaZulu-Natal, South Africa (The word *Kloof* means 'gorge' in Afrikaans)
- Komati Gorge, South Africa
- Lanner Gorge, South Africa
- Mides Canyon, Tunisia
- Olduvai Gorge, Great Rift Valley, Africa
- Talari Gorges, Mali

Americas

- Atuel Canyon, Mendoza Province, Argentina
- Bad Rock Canyon, Montana, United States
- Big South Fork of the Cumberland River gorge, Kentucky and Tennessee
- Black Canyon of the Gunnison, Colorado, United States
- Breaks Canyon. Kentucky/Virginia, United States
- Canyon de Chelly, Arizona, United States
- Canyon Sainte-Anne, Quebec, Canada
- Chicamocha Canyon, Santander, Colombia
- Colca Canyon, Arequipa, Peru
- Columbia River Gorge, Washington and Oregon, United States
- Copper Canyon, Chihuahua, Mexico
- Cotahuasi Canyon, Arequipa, Peru
- Farwell Canyon, British Columbia, Canada
- Flume Gorge, New Hampshire, United States
- Fraser Canyon, British Columbia, Canada
- Grand Canyon, Arizona, United States
- Grand Canyon of the Stikine, British Columbia, Canada

- Grand Canyon of the Yellowstone, Wyoming, United States
- Grand Canyon of Torotoro, Torotoro, Potosi Department, Bolivia
- Guartelá Canyon, Paraná, Brazil
- Hells Canyon, Idaho and Oregon, United States
- Horseshoe Canyon, Alberta, Canada
- Huasteca Canyon, Monterrey, Mexico
- Itaimbezinho, Aparados da Serra National Park, Rio Grande do Sul, Brazi
- James River Gorge, Virginia, United States
- Kings Canyon National Park, California, United States
- Lehigh River Gorge, The Poconos, Northeastern Pennsylvania
- Letchworth State Park (Genesee River Gorge), New York, United States
- Linville Gorge Wilderness, North Carolina, United States
- Little River Canyon, Alabama, United States
- Logan Canyon, Utah, United States
- New River Gorge, West Virginia, United States
- Niagara Gorge, Canada/United States
- Nine Mile Canyon, Utah, United States
- Ouimet Canyon, Ontario, Canada
- Palo Duro Canyon, Texas, United States
- Painted Chasm, Chasm Provincial Park, British Columbia, Canada
- Pine Creek Gorge, Pennsylvania, United States
- Quechee Gorge, Vermont, United States
- Red River Gorge, Kentucky, United States
- Rio Grande Gorge, New Mexico, United States
- Ripogenus Gorge, Maine, United States
- Royal Gorge, Colorado, United States
- Saturban canyon, Santander, Colombia
- Somoto Canyon, Somoto, Madriz, Nicaragua

- St. Christopher's Canyon (Cañon de San Cristobal), Barranquitas/Aibonito, Puerto Rico
- Sumidero Canyon, Chiapas, Mexico
- Tallulah Gorge, Georgia, United States
- Waimea Canyon, Hawaii, United States

Asia

- Ala Archa gorge of Ala Archa National Park, Kyrgyz Republic
- Bued Gorge, Benguet, Philippines
- Charyn Canyon, Kazakhstan
- Chovar Gorge, Kathmandu, Nepal
- Cukang Taneuh, Pangandaran, Indonesia
- Indus River Gorge through the Himalaya, Pakistan
- Kali Gandaki Gorge, Gandaki, Nepal
- Montalban Gorge, Philippines
- Siq, Petra, Jordan
- Soan River george, Pakistan
- Taroko Gorge of Taroko National Park, Taiwan
- Three Gorges, Chongqing, China
- Tiger Leaping Gorge, Yunnan, China
- Yarlung Zangbo Grand Canyon, Tibet Autonomous Region, China

Europe

- Aare Gorge, Kanton Bern, Switzerland
- Aragvi River Gorge, Georgia
- Ardèche Gorges, Rhône-Alpes, France
- Avon Gorge, Bristol, England, UK
- Bletterbach, South Tyrol, Italy
- Cheddar Gorge, Somerset, England, UK
- Corrieshalloch Gorge, Ullapool, Scotland, UK

- Daluis Gorge, Provence, France
- Dniester Canyon, Ukraine
- Valla Canyon, Kure, Kastamonu, Turkey
- Ebbor Gorge, Somerset, England, UK
- Fjaðrárgljúfur Canyon, Iceland
- Shnizow Canyon, Uşak, Turkey
- Gorges du Tarn, France
- Gorropu, Sardinia, Italy
- Greenland's Grand Canyon, Greenland
- Ironbridge Gorge, Shropshire, England, UK
- Osum Gorge, Albania
- Liechtensteinklamm, Salzburg (state), Austria
- Matka Canyon, Republic of Macedonia
- Pankisi Gorge, Georgia
- Richtis Gorge, Crete, Greece
- Rugova Canyon, Serbia
- Samaria Gorge, Crete, Greece
- Sautso, Finnmark, Norway
- Tara River Canyon, Montenegro
- Harmanköy Canyon, Bilecik, Turkey
- Verdon Gorge, Provence, France
- Viamala, Graubünden, Switzerland
- Vikos Gorge, Vikos–Aoös National Park, Greece

Oceania

Australia

- Barfold Gorge, Victoria
- Barron Gorge, Queensland
- Bouldercombe Gorge, Queensland

- Cambanoora Gorge, Queensland
- Capertee Valley, New South Wales
- Carnarvon Gorge, Queensland
- Cataract Gorge, Tasmania
- Dimond Gorge, Western Australia
- Galston Gorge, New South Wales
- Katherine Gorge, Northern Territory
- Kings Canyon, Northern Territory
- Lerderderg Gorge, Victoria
- Loch Ard Gorge, Victoria
- Little River Gorge, Victoria
- Mossman Gorge, Queensland
- Murchison River Gorge, Western Australia
- Nepean Gorge, New South Wales
- North and South Gorges of North Stradbroke Island, Queensland
- Porcupine Gorge, Queensland
- Palm Valley, Northern Territory
- Ravine des Casoars, South Australia
- Sturt Gorge, South Australia
- Windjana Gorge, Western Australia

New Zealand

- Buller Gorge, Buller, New Zealand
- Manawatu Gorge, Manawatu, North Island, New Zealand
- Skippers Canyon, New Zealand

Middle East

- Wadi Ghul, Oman

List of other Features Causing Gorges or Canyons

- Ardèche River, France
- Bued River, Philippines
- Danube River, Europe
- Hérault River, France
- Iron Gate, Serbia-Romania
- Kings River of the Kings Canyon National Park, California, United States
- Ruinaulta, cutting Flims Rockslide, Switzerland
- South Nahanni River in Canada
- Sierra y Cañones de Guara, Alquezar, Spain
- Steall Waterfall of Glen Nevis, Lochaber, Scotland, UK

Canyons on Other Planetary Bodies

- Ithaca Chasma on Saturn's moon Tethys
- Valles Marineris on Mars, the largest known canyon in the solar system
- Vid Flumina on Saturn's largest moon Titan is the only known liquid-floored canyon in the solar system besides Earth

Venus has many craters and canyons on its surface. The troughs on the planet are part of a system of canyons that is more than 6 400 km long.

Valley

A valley is a low area between hills, often with a river running through it.

Calchaquí Valleys in Argentina

In geology, a valley or dale is a depression that is longer than it is wide. The terms U-shaped and V-shaped are descriptive terms of geography to characterize the form of valleys. Most valleys belong to one of these two main types or a mixture of them, (at least) with respect to the cross section of the slopes or hillsides.

U-shaped valley in Glacier National Park, Montana, United States

Romsdalen in Western Norway is an almost vertical valley.

Fljótsdalur in East Iceland, a rather flat valley (in Scotland, this type of valley is called a "strath")

The Frades Valley in the mountainous region of Rio de Janeiro state, Brazil

Valley Terminology

A valley in its broadest geographic sense is also known as a *dale*. A valley through which a river runs may also be referred to as a vale. A small, secluded, and often wooded valley is known as a dell or in Scotland as a glen. A wide, flat valley through which a river runs is known in Scotland as a strath. A mountain cove is a small valley, closed at one or both ends, in the central or southern Appalachian Mountains which sometimes results from the erosion of a geologic window. A small valley surrounded by mountains or ridges is sometimes known as a hollow. A deep, narrow valley is known as a cwm (also spelled combe or coombe). Similar geological structures, such as canyons, ravines, gorges, gul-lies, and kloofs, are not usually referred to as valleys. A valley formed by erosion is called an erosional valley; a valley formed by geologic events such as drop faults or the rise of highlands is called a structural valley.

River Valleys

A valley formed by flowing water, or *river valley*, is usually V-shaped. The exact shape will depend on the characteristics of the stream flowing through it. Rivers with steep gradients, as in mountain ranges, produce steep walls and a bottom. Shallower slopes may produce broader and gentler valleys. However, in the lowest stretch of a river, where it approaches its base level, it begins to deposit sediment and the valley bottom becomes a floodplain.

Some broad *V* examples are:

- North America: Black Canyon of the Gunnison National Park, and others in Grand Canyon NP
- Europe:
    - Austria: narrow passages of upper Inn valley (Inntal), affluents of Enns
    - Switzerland: Napf region, Zurich Oberland, Engadin
    - Germany: affluents to the middle reaches of Rhine and Mosel

Some of the first human complex societies originated in river valleys, such as that of the Nile, Tigris-Euphrates, Indus, Ganges, Yangtze, Yellow River, Mississippi, and arguably Amazon. In prehistory, the rivers were used as a source of fresh water and food (fish and game), as well as a place to wash and a sewer. The proximity of water moderated temperature extremes and provided a source for irrigation, stimulating the development of agriculture. Most of the first civilizations developed from these river valley communities.

Vale

In geography, a *vale* is a wide river valley, usually with a particularly wide flood plain or flat valley bottom.

In Southern England, vales commonly occur between the escarpment slopes of pairs of chalk formations, where the chalk dome has been eroded, exposing less resistant underlying rock, usually claystone.

Glacial Valleys

U-shaped valley on the Afon Fathew near Dolgoch, Wales

Yosemite Valley from an airplane

A glaciated valley in the Mount Hood Wilderness showing a characteristic U-shape, the bottom's rocky 'rubble' accretion and the broad shoulders

There are various forms of valley associated with glaciation that may be referred to as glacial valleys.

U-shaped Valley

A valley carved by glaciers is normally U-shaped. The valley becomes visible upon the recession of the glacier that forms it. When the ice recedes or thaws, the valley remains, often littered with small boulders that were transported within the ice. Floor gradient does not affect the valley's shape, it is the glacier's size that does. Continuously flowing glaciers – especially in the ice age – and large-sized glaciers carve wide, deep incised valleys.

Examples of U-shaped valleys are found in every mountainous region that has experienced glaciation, usually during the Pleistocene ice ages. Most present U-shaped valleys started as V-shaped before glaciation. The glaciers carved it out wider and deeper, simultaneously changing the shape. This proceeds through the glacial erosion processes of glaciation and abrasion, which results in large rocky material (glacial till) being carried in the glacier. A material called boulder clay is deposited on the floor of the valley. As the ice melts and retreats, the valley is left with very steep sides and a wide, flat floor. A river or stream may remain in the valley. This replaces the original stream or river and is known as a misfit stream because it is smaller than one would expect given the size of its valley.

Other interesting glacially carved valleys include:

- Yosemite Valley (United States)
- Side valleys of the Austrian river Salzach for their parallel directions and hanging mouths.
- Some Scottish glens full with bushes and flowers.
- That of the St. Mary River in Glacier National Park in Montana, USA.

Tunnel Valley

A tunnel valley is a large, long, U-shaped valley originally cut under the glacial ice near the margin of continental ice sheets such as that now covering Antarctica and formerly covering portions of all continents during past glacial ages.

A tunnel valley can be up to 100 km (62 mi) long, 4 km (2.5 mi) wide, and 400 m (1,300 ft) deep (its depth may vary along its length).

Tunnel valleys were formed by subglacial erosion by water. They served as subglacial drainage pathways carrying large volumes of melt water. Their cross-sections exhibit steep-sided flanks similar to fjord walls, and their flat bottoms are typical of subglacial glacial erosion.

Meltwater Valley

In northern Central Europe, the Scandinavian ice sheet during the various ice ages advanced slightly uphill against the lie of the land. As a result, its meltwaters flowed parallel to the ice margin to reach the North Sea basin, forming huge, flat valleys known as *Urstromtäler*. Unlike the other forms of glacial valley, these were formed by glacial meltwaters.

Transition Forms and Valley Shoulders

Depending on the topography, the rock types and the climate, a lot of transitional forms between V-, U- and plain valleys exist. Their bottoms can be broad or narrow, but characteristic is also the type of valley shoulder. The broader a mountain valley, the lower its shoulders are located in most cases. An important exception are canyons where the shoulder almost is near the top of the valley's slope. In the Alps – e.g. the Tyrolean Inn valley – the shoulders are quite low (100–200 meters above the bottom). Many villages are located here (esp. at the sunny side) because the climate is very mild: even in winter when the valley's floor is completely filled with fog, these villages are in sunshine.

Look from Paria View to a valley in Bryce Canyon, Utah, with very striking shoulders

In some stress-tectonic regions of the Rockies or the Alps (e.g. Salzburg) the side valleys are parallel to each other, and additionally they are hanging. The brooks flow into the river in form of deep canyons or waterfalls. Usually this fact is the result of a violent erosion of the former valley shoulders, a special genesis found also at arêtes and glacial cirques, at every Scottish glen, or in a northern fjord.

Hanging Valleys

Bridal Veil Falls in Yosemite National Park flowing from a hanging valley.

A hanging valley is a tributary valley that is higher than the main valley. They are most commonly associated with U-shaped valleys when a tributary glacier flows into a glacier of larger volume. The main glacier erodes a deep U-shaped valley with nearly vertical sides while the tributary glacier, with a smaller volume of ice, makes a shallower U-shaped valley. Since the surfaces of the glaciers were originally at the same elevation, the shallower valley appears to be 'hanging' above the main valley. Often, waterfalls form at or near the outlet of the upper valley.

Glaciated terrain is not the only site of hanging streams and valleys. Hanging valleys are also simply the product of varying rates of erosion of the main valley and the tributary valleys. The varying rates of erosion are associated with the composition of the adjacent rocks in the different valley locations. The tributary valleys are eroded and deepened by glaciers or erosion at a slower rate than that of the main valley floor, thus the difference in the two valleys' depth increases over time. The tributary valley composed of more resistant rock then hangs over the main valley.

Valley Floors

Usually the bottom of a main valley is broad – independent of the U or V shape. It typically ranges from about one to ten kilometers in width and is commonly filled with mountain sediments. The shape of the floor can be rather horizontal, similar to a flat cylinder, or terraced.

Side valleys are rather V than U-shaped; near the mouth waterfalls are possible if it is a hanging valley. The location of the villages depends on the across-valley profile, on climate and local traditions, and on the danger of avalanches or landslides. Predominant are places on terraces or alluvial fans if they exist.

Historic siting of villages within the mainstem valleys, however, have chiefly considered the potential of flooding.

Hollows

Wheat in the Hula Valley, Israel

River Indus running through Kohistan valley in Pakistan

A hollow (holler) is a small valley or dry stream bed. This term is commonly used in New England, Appalachia, Ozarks, Arkansas and Missouri to describe such geographic features. In rural areas, it may be pronounced as "holler". Hollows may be formed by river valleys, such as Mansfield Hollow, or they may be relatively dry clefts with a notch-like characteristic in that they have a height of land and consequent water divide in their bases.

Famous Valleys

The Valley of Flowers in Uttarakhand, India

The Harau Valley in West Sumatra, Indonesia

Hell's Gate, British Columbia

A view of Orosí, Costa Rica

- Aburra Valley (Colombia)
- Barossa Valley (Australia)
- Cauca Valley (Colombia)
- Central Valley (California) (California, United States)
- Dalen, Telemark (Telemark, Norway)
- Dang Valley (Western Nepal)
- Danube Valley (Eastern Europe)
- Death Valley (California, United States)
- Fraser Canyon (British Columbia, Canada)
- Fraser Valley (British Columbia, Canada)
- Glen Coe (Scotland, United Kingdom)
- Grand Canyon (Arizona, United States)
- Great Glen (Scotland, United Kingdom)
- Gudbrand Valley (Oppland, Norway)
- Hallingdalen (Buskerud, Norway)
- Hell's Gate (British Columbia, Canada)
- Heddal (Telemark, Norway)
- Hudson Valley (New York, United States)
- Hunter Region (Australia)
- Hunza Valley (Pakistan)
- Hutt Valley (New Zealand)
- Ihlara, Turkey
- Indus Valley (Pakistan)
- Iron Gate (Romania/Serbia)
- Jordan Valley
- Kaghan Valley (Pakistan)

- Kathmandu (Nepal)
- Las Vegas Valley (Nevada, United States)
- Lauterbrunnen Valley (Bern, Switzerland)
- Little Cottonwood Creek Valley (Utah, United States)
- Loire Valley with its famous castles (France)
- Valley of the Moon, San Juan Argentina
- Midt-Telemark (Telemark, Norway)
- Monument Valley (United States)
- Nant Ffrancon (Wales, United Kingdom)
- Napa Valley (California, United States)
- Neelum Valley (Pakistan)
- Nile Valley (Egypt/Sudan/Ethiopia/Uganda)
- Numedalen (Buskerud, Norway)
- Okanagan Valley (British Columbia, Canada)
- Ottawa Valley (Ontario/Quebec, Canada)
- Palo Duro Canyon (Texas, United States)
- Panjshir Valley (Afghanistan)
- Upper Rhine Valley or Upper Rhine Plain, an old graben system. (France and Germany)
- Rhone Valley from the Matterhorn to Grenoble and Lyon (France)
- Rio Grande Valley (Texas, United States)
- Romsdalen (Møre Og Romsdal, Norway)
- Saint Lawrence Valley, the most densely populated region of Canada. (Ontario/Quebec/New York) (Canada/United States)
- San Fernando Valley (California, United States)
- Setesdal (Agder, Norway)
- Shenandoah Valley (Virginia/West Virginia, United States)
- Silicon Valley, (California, United States)

- Sonoma Valley, (California, United States)
- South Wales Valleys (Wales, United Kingdom)
- Swat Valley (Pakistan)
- Valley of flowers (India)
- Valley of the Gods (United States)
- Valley of the Kings (Egypt)
- Valley of Mexico (Mexico), also known as "El Valle de México", Mexico City
- Valley of the Sun (Arizona, United States)
- Vestfjorddalen (Norway)
- Willamette Valley, (Oregon, United States)
- Jordan Rift Valley (Jordan - Israel)
- Østerdalen (Hedmark, Norway)

Rift Valleys

Rift valleys, such as the Albertine Rift, are formed by the expansion of the Earth's crust due to tectonic activity beneath the Earth's surface.

Mid-ocean Ridge

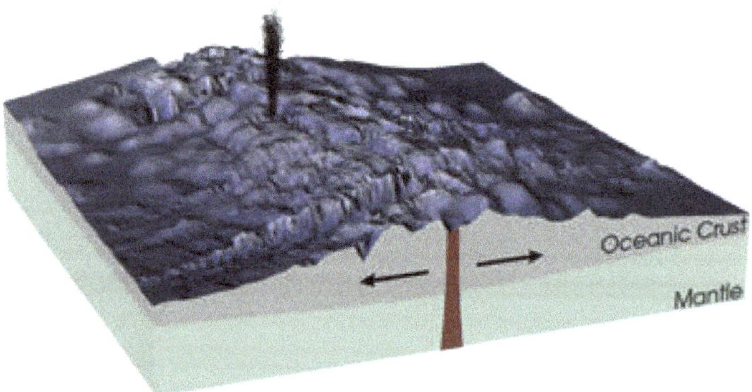

Mid-oceanic ridge, including a black smoker

A mid-ocean ridge is an underwater mountain system formed by plate tectonics. It consists of various mountains linked in chains, typically having a valley known as a rift running along its spine. This type of oceanic mountain ridge is characteristic of what is known as an oceanic spreading center, which is responsible for seafloor spreading. The production of new seafloor results from mantle upwelling in response to plate spreading; this isentropic upwelling solid mantle material eventually exceeds the solidus and melts. The buoyant melt rises as magma at a linear weakness in the oceanic crust, and emerges as lava, creating new crust upon cooling. A mid-ocean ridge demarcates the boundary between two tectonic plates, and consequently is termed a divergent plate boundary.

Description

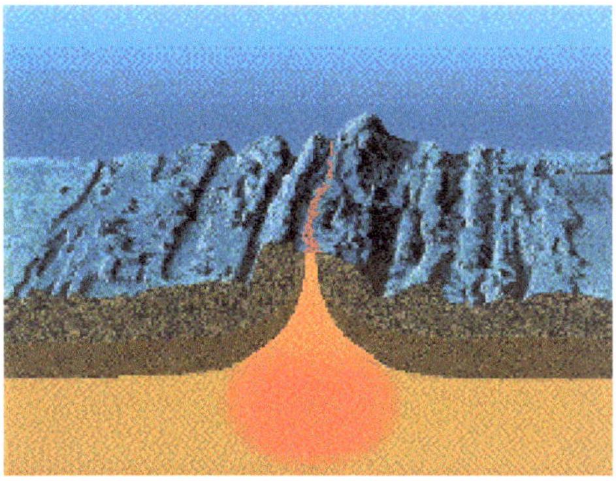

A close-up showing a mid-ocean ridge topography with magma rising from a chamber below, forming new ocean plate which spreads away from ridge

Mid-ocean ridges are geologically active, with new magma constantly emerging onto the ocean floor and into the crust at and near rifts along the ridge axes. The crystallized magma forms new crust of basalt (known as MORB for mid-ocean ridge basalt) and gabbro. They are formed by two oceanic plates moving away from each other.

The rocks making up the crust below the seafloor are youngest at the axis of the ridge and age with increasing distance from that axis. New magma of basalt composition emerges at and near the axis because of decompression melting in the underlying Earth's mantle.

The oceanic crust is made up of rocks much younger than the Earth itself. Most oceanic crust in the ocean basins is less than 200 million years old. The crust is in a constant state of "renewal" at the ocean ridges. Moving away from the mid-ocean ridge, ocean depth progressively increases; the greatest depths are in ocean trenches. As the oceanic crust moves away from the ridge axis, the peridotite in the underlying mantle cools and becomes more rigid. The crust and the relatively rigid peridotite below it make up the oceanic lithosphere.

Slow spreading ridges like the Mid-Atlantic Ridge (MAR) generally have large, wide rift valleys, sometimes as wide as 10–20 km (6.2–12.4 mi), and very rugged terrain at the ridge crest that can have relief of up to a 1,000 m (3,300 ft). By contrast, fast spreading ridges like the East Pacific Rise (EPR) are narrow, sharp incisions surrounded by generally flat topography that slopes away from the ridge over many hundreds of miles.

The overall shape of ridges results from Pratt isostacy: close to the ridge axis there is hot, low-density mantle supporting the oceanic crust. As the oceanic plates cool, away from the ridge axes, the oceanic mantle lithosphere (the colder, denser part of the mantle that, together with the crust, comprises the oceanic plates) thickens and the density increases. Thus older seafloor is underlain by denser material and 'sits' lower. The width of the ridge is hence a function of spreading rate - slow ridges like the MAR have spread much less far than faster ridges like the EPR for the same amount of cooling and consequent bathymetric drop-off.

Formation Processes

There are two processes, ridge-push and slab pull, thought to be responsible for the spreading seen at mid-ocean ridges, and there is some uncertainty as to which is dominant. Ridge-push occurs when the growing bulk of the ridge pushes the rest of the tectonic plate away from the ridge, often towards a subduction zone. At the subduction zone, "slab-pull" comes into effect. This is simply the weight of the tectonic plate being subducted (pulled) below the overlying plate dragging the rest of the plate along behind it.

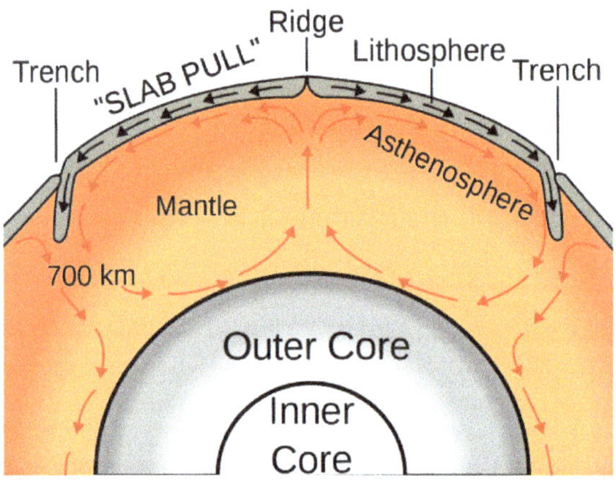

Oceanic crust is formed at an oceanic ridge, while the lithosphere is subducted back into the asthenosphere at trenches.

The other process proposed to contribute to the formation of new oceanic crust at mid-ocean ridges is the "mantle conveyor" (see image). However, there have been some studies which have shown that the upper mantle (asthenosphere) is too plastic (flexible) to generate enough friction to pull the tectonic plate along. Moreover, unlike in the image above, mantle upwelling that causes magma to form beneath the ocean ridges appears to involve only its upper 400 km (250 mi), as deduced from seismic tomography and from studies of the seismic discontinuity at about 400 km (250 mi). The relatively shallow depths from which the upwelling mantle rises below ridges are more consistent with the "slab-pull" process. On the other hand, some of the world's largest tectonic plates such as the North American Plate are in motion, yet are nowhere being subducted.

The rate at which the mid-ocean ridge creates new material is known as the spreading rate, and is generally measured in mm/yr. The common subdivisions of spreading rate are fast, medium, and slow with values generally being >100 mm/yr, 100–55 mm/yr, and 55–20 mm/yr, respectively. The spreading rate of the North Atlantic Ocean is ~ 25 mm/yr, while in the Pacific region, it is 80–120 mm/yr. Ridges that spread at rates <20 mm/yr are referred to as ultraslow spreading ridges (e.g., the Gakkel Ridge in the Arctic Ocean and the Southwest Indian Ridge) and they provide a much different perspective on crustal formation than their faster spreading brethren.

The mid-ocean ridge systems form new oceanic crust. As crystallized basalt extruded at a ridge axis cools below Curie points of appropriate iron-titanium oxides, magnetic field directions parallel to the Earth's magnetic field are recorded in those oxides. The orientations of the field in the oceanic crust record preserve a record of directions of the Earth's magnetic field with time. Because the field has reversed directions at irregular intervals throughout its history, the pattern of geomagnetic reversals in the ocean crust can be used as an indicator of age. Likewise, the pattern of reversals together with age measurements of the crust is used to help establish the history of the Earth's magnetic field.

Global System

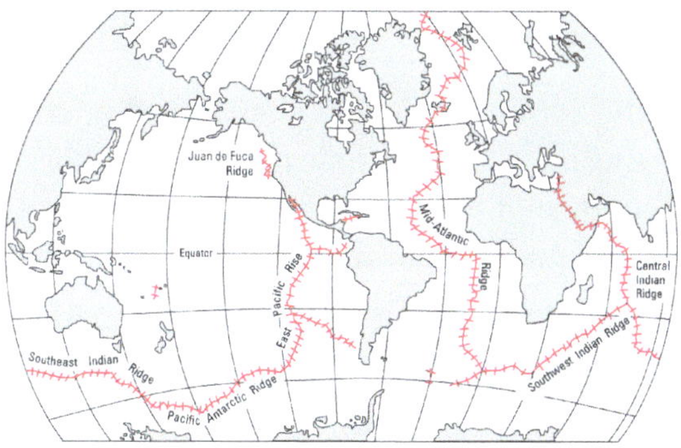

World Distribution of Mid-Oceanic Ridges; USGS

The mid-ocean ridges of the world are connected and form *the* Ocean Ridge, a single global mid-oceanic ridge system that is part of every ocean, making it the longest mountain range in the world. The continuous mountain range is 65,000 km (40,400 mi) long (several times longer than the Andes, the longest continental mountain range), and the total length of the oceanic ridge system is 80,000 km (49,700 mi) long.

History

Discovery

Mid-ocean ridges are generally submerged deep in the ocean. It was not until the 1950s, when the ocean floor was surveyed in detail, that their full extent became known.

The *Vema*, a ship of the Lamont-Doherty Earth Observatory of Columbia University, traversed the Atlantic Ocean, recording data about the ocean floor from the ocean surface. A team led by Marie Tharp and Bruce Heezen analyzed the data and concluded that there was an enormous mountain chain running up the middle. Scientists named it the "Mid-Atlantic Ridge".

At first, the ridge was thought to be a phenomenon specific to the Atlantic Ocean. However, as surveys of the ocean floor continued around the world, it was discovered that every ocean contains parts of the mid-ocean ridge system. Although the ridge system runs down the middle of the Atlantic Ocean, the ridge is located away from the center of other oceans.

Impact

Alfred Wegener proposed the theory of continental drift in 1912. He stated: "the Mid-Atlantic Ridge ... zone in which the floor of the Atlantic, as it keeps spreading, is continuously tearing open and making space for fresh, relatively fluid and hot sima [rising] from depth." However, Wegener did not pursue this observation in his later works and his theory was dismissed by geologists because there was no mechanism to explain how continents could plow through ocean crust, and the theory became largely forgotten.

Following the discovery of the worldwide extent of the mid-ocean ridge in the 1950s, geologists faced a new task: explaining how such an enormous geological structure could have formed. In the 1960s, geologists discovered and began to propose mechanisms for seafloor spreading. Plate tectonics was a suitable explanation for seafloor spreading, and the acceptance of plate tectonics by the majority of geologists resulted in a major paradigm shift in geological thinking.

It is estimated that 20 volcanic eruptions occur each year along earth's mid-ocean ridges and that every year 2.5 km² (0.97 sq mi) of new seafloor is formed by this process. With a crustal thickness of 1 to 2 km (0.62 to 1.24 mi), this amounts to about 4 km³ (0.96 cu mi) of new ocean crust formed every year.

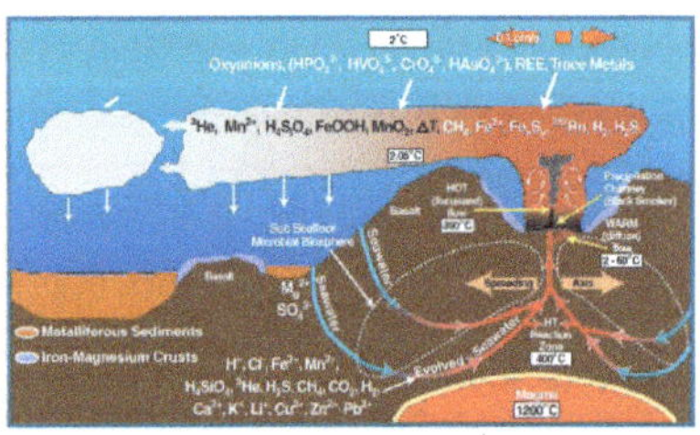

Oceanic ridge and deep sea vent chemistry

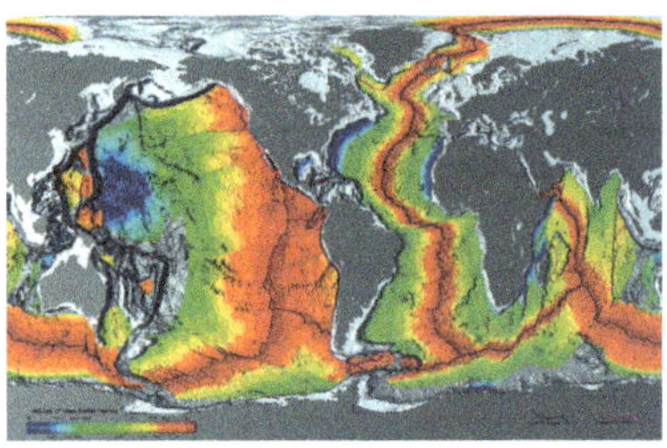

Age of oceanic crust. The red is most recent, and blue is the oldest.

List of Oceanic Ridges

- Aden Ridge
- Cocos Ridge
- Explorer Ridge
- Gorda Ridge
- Juan de Fuca Ridge
- American-Antarctic Ridge
- Chile Rise
- East Pacific Rise
- East Scotia Ridge
- Gakkel Ridge (Mid-Arctic Ridge)
- Nazca Ridge
- Pacific-Antarctic Ridge
- Central Indian Ridge
    - Carlsberg Ridge
- Southeast Indian Ridge
- Southwest Indian Ridge
- Mid-Atlantic Ridge
    - Kolbeinsey Ridge (North of Iceland)
    - Mohns Ridge
    - Knipovich Ridge (between Greenland and Spitsbergen)
    - Reykjanes Ridge (South of Iceland)

List of Ancient Oceanic Ridges

- Aegir Ridge
- Alpha Ridge
- Kula-Farallon Ridge

- Pacific-Farallon Ridge
- Pacific-Kula Ridge
- Phoenix Ridge

References

- Whittow, John (1984). Dictionary of Physical Geography. London: Penguin, 2004, p. 352. ISBN 0-14-051094-X.
- Nuttall, John & Anne (2008). England. The Mountains of England & Wales. 2 (3rd ed.). Milnthorpe, Cumbria: Cicerone. ISBN 1-85284-037-4.
- Helman, Adam (2005). The Finest Peaks - Prominence and Other Mountain Measures. Trafford Publishing. ISBN 978-1412059954.
- "Mountain building". Science matters: earth and beyond; module 4. Pearson South Africa. 2002. p. 75. ISBN 0-7986-6059-7.
- Butz, Stephen D (2004). "Chapter 8: Plate tectonics". Science of Earth Systems. Thompson/Delmar Learning. p. 136. ISBN 0-7668-3391-7.
- Tweit, Susan J. (1992). The Great Southwest Nature Factbook. Alaska Northwest Books. pp. 209–210. ISBN 0-88240-434-2.
- Fraknoi, A.; Morrison, D.; Wolff, S. (2004). Voyages to the Planets (3rd ed.). Belmont: Thomson Books/Cole. ISBN 9780534395674.
- Gerrard, A.J. (1990). Mountain Environments: An Examination of the Physical Geography of Mountains. Cambridge, Mass: MIT Press. ISBN 978-0262071284.
- Cohen, Callan & Spottiswoode, Claire & Rossouw, Jonathan (2006). Southern African Birdfinder. p. 210. ISBN 1-86872-725-4.
- Cox, Steven M.; Fulsaas, Kris, eds. (2009) [2003]. Mountaineering: The Freedom of the Hills (7 ed.). Seattle: The Mountaineers. ISBN 0-89886-828-9.
- Fitzsimons, David (14 December 2015). "Capertee Valley: Australia's own Grand Canyon". Daily Telegraph. Retrieved March 28, 2016.
- Kruszelnicki, Dr. Karl S. (22 May 2012). "Grand Canyon is not so grand › Dr Karl's Great Moments In Science". ABC Science. Retrieved March 28, 2016.
- "What is the difference between lake and pond; mountain and hill; or river and creek?". USGS. Archived from the original on 2013-02-12. Retrieved 11 February 2013.
- "Biotic Communities of the Colorado Plateau: C. Hart Merriam and the Life Zones Concept". Archived from the original on 2013-02-04. Retrieved 30 January 2010.

4

Techniques of Topography

Surveying is a method that helps in the determination of terrestrial positioning of the points and the distances between them. These points are usually used to establish land maps and boundaries. Like surveying the other techniques of topography are remote sensing, satellite imagery, topographic prominence, aerial photography, bathymetry, geovisualization etc. This text examines the methods of topography in a critical manner providing key analysis to the subject matter.

Surveying

Surveying or land surveying is the technique, profession, and science of determining the terrestrial or three-dimensional position of points and the distances and angles between them. A land surveying professional is called a land surveyor. These points are usually on the surface of the Earth, and they are often used to establish land maps and boundaries for ownership, locations like building corners or the surface location of subsurface features, or other purposes required by government or civil law, such as property sales.

A surveyor at work with an infrared reflector used for distance measurement.

Surveyors work with elements of geometry, trigonometry, regression analysis, physics, engineering, metrology, programming languages and the law. They use equipment like total stations, robotic total stations, GPS receivers, retroreflectors, 3D scanners, radios, handheld tablets, digital levels, drones, GIS and surveying software.

Surveying has been an element in the development of the human environment since the beginning of recorded history. The planning and execution of most forms of construction require it. It is also used in transport, communications, mapping, and the definition of legal boundaries for land ownership. It is an important tool for research in many other scientific disciplines.

Definitions

ACSM

The American Congress on Surveying and Mapping (ACSM), defines surveying as *the science and art of making all essential measurements to determine the relative position of points or physical and cultural details above, on, or beneath the surface of the Earth, and to depict them in a usable form, or to establish the position of points or details.*

Also per ACSM, the type of surveying known as "land surveying" is *the detailed study or inspection, as by gathering information through observations, measurements in the field, questionnaires, or research of legal instruments, and data analysis in the support of planning, designing, and establishing of property boundaries. It involves the re-establishment of cadastral surveys and land boundaries based on documents of record and historical evidence, as well as certifying surveys (as required by statute or local ordinance) of subdivision plats or maps, registered land surveys, judicial surveys, and space delineation. Land surveying can include associated services such as mapping and related data accumulation, construction layout surveys, precision measurements of length, angle, elevation, area, and volume, as well as horizontal and vertical control surveys, and the analysis and utilization of land survey data.*

FIG

The International Federation of Surveyors defines the function of surveying as:

A surveyor is a professional person with the academic qualifications and technical expertise to conduct one, or more, of the following activities;

- to determine, measure and represent land, three-dimensional objects, point-fields and trajectories;
- to assemble and interpret land and geographically related information,
- to use that information for the planning and efficient administration of the land, the sea and any structures thereon; and,
- to conduct research into the above practices and to develop them.

History

Ancient Surveying

Basic surveyance has occurred since humans built the first large structures. The prehistoric monument at Stonehenge (c. 2500 BC) was set out by prehistoric surveyors using peg and rope geometry.

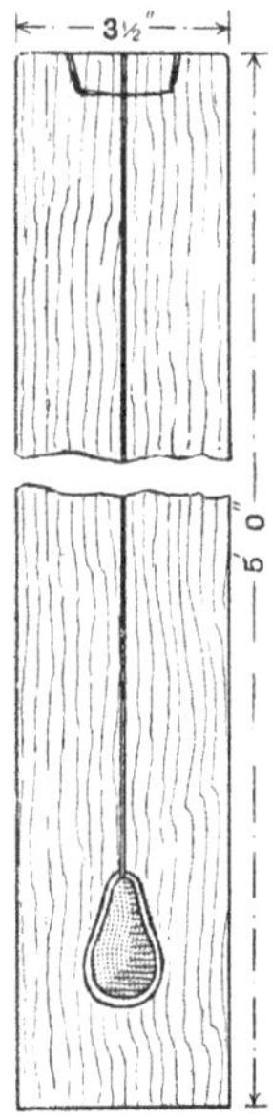

A plumb rule from the book Cassells' Carpentry and Joinery

In ancient Egypt, a rope stretcher would use simple geometry to re-establish boundaries after the annual floods of the Nile River. The almost perfect squareness and north-south orientation of the Great Pyramid of Giza, built c. 2700 BC, affirm the Egyptians' command of surveying. The *Groma* instrument originated in Mesopotamia (early 1st millennium BC).

The mathematician Liu Hui described ways of measuring distant objects in his work *Haidao suanjing* or *The Sea Island Mathematical Manual*, published in 263 AD.

The Romans recognized land surveyors as a profession. They established the basic measurements under which the Roman Empire was divided, such as a tax register of conquered lands (300 AD). Roman surveyors were known as *Gromatici*.

In medieval Europe, beating the bounds maintained the boundaries of a village or parish. This was the practice of gathering a group of residents and walking around the parish or village to establish a communal memory of the boundaries. Young boys were included to ensure the memory lasted as long as possible.

In England, William the Conqueror commissioned the Domesday Book in 1086. It recorded the names of all the land owners, the area of land they owned, the quality of the land, and specific information of the area's content and inhabitants. It did not include maps showing exact locations.

Modern Surveying

Abel Foullon described a plane table in 1551, but it is thought that the instrument was in use earlier as his description is of a developed instrument.

Gunter's chain was introduced in 1620 by English mathematician Edmund Gunter. It enabled plots of land to be accurately surveyed and plotted for legal and commercial purposes.

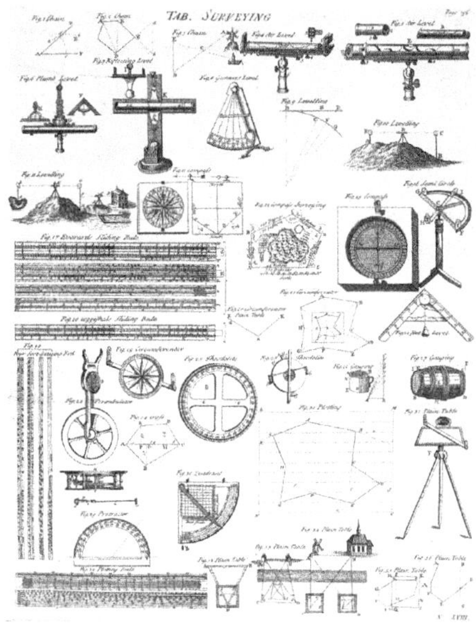

Table of Surveying, 1728 *Cyclopaedia*

Leonard Digges described a Theodolite that measured horizontal angles in his book *A geometric practice named Pantometria* (1571). Joshua Habermel (de:Erasmus Habermehl) created a theodolite with a compass and tripod in 1576. Johnathon Sission was the first to incorporate a telescope on a theodolite in 1725.

In the 18th century, modern techniques and instruments for surveying began to be used. Jesse Ramsden introduced the first precision theodolite in 1787. It was an instrument for measuring angles in the horizontal and vertical planes. He created his great theodolite using an accurate dividing engine of his own design. Ramsden's theodolite represented a great step forward in the instrument's accuracy. William Gascoigne invented an instrument that used a telescope with an installed crosshair as a target device, in 1640. James Watt developed an optical meter for the measuring of distance in 1771; it measured the parallactic angle from which the distance to a point could be deduced.

Dutch mathematician Willebrord Snellius (a.k.a. Snell) introduced the modern systematic use of triangulation. In 1615 he surveyed the distance from Alkmaar to Bergen op Zoom, approximately 70 miles (110 kilometres). The survey was a chain of quadrangles containing 33 triangles in all. Snell showed how planar formulae could be corrected to allow for the curvature of the earth. He also showed how to resection, or calculate, the position of a point inside a triangle using the angles cast between the vertices at the unknown point. These could be measured more accurately than bearings of the vertices, which depended on a compass. His work established the idea of surveying a primary network of control points, and locating subsidiary points inside the primary network later. Between 1733 and 1740, Jacques Cassini and his son César undertook the first triangulation of France. They included a re-surveying of the meridian arc, leading to the publication in 1745 of the first map of France constructed on rigorous principles. By this time, triangulation methods were by then well established for local map-making.

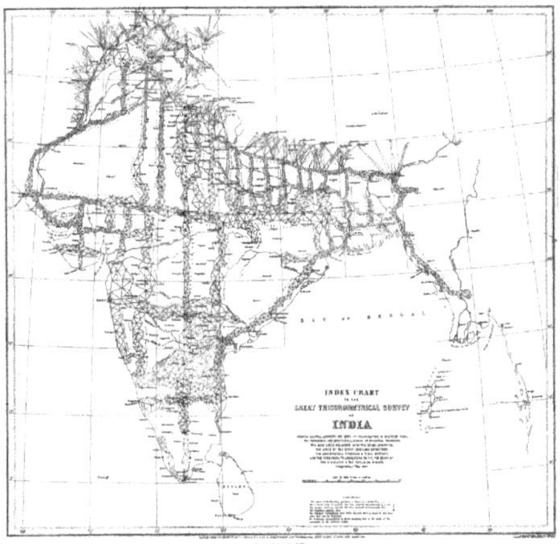

A map of India showing the Great Trigonometrical Survey, produced in 1870

It was only towards the end of the 18th century that detailed triangulation network surveys mapped whole countries. In 1784, a team from General William Roy's Ordnance Survey of Great Britain began the Principal Triangulation of Britain. The first Ramsden theodolite was built for this survey. The survey was finally completed in 1853. The Great Trigonometric Survey of India began in 1801. The Indian survey had an enormous scientific impact. It was responsible for one of the first accurate measurements of a section of an arc of longitude, and for measurements of the geodesic anomaly. It named and mapped Mount Everest and the other Himalayan peaks. Surveying became a professional occupation in high demand at the turn of the 19th century with the onset of the Industrial Revolution. The profession developed more accurate instruments to aid its work. Industrial infrastructure projects used surveyors to lay out canals, roads and rail.

In the US, the Land Ordinance of 1785 created the Public Land Survey System. It formed the basis for dividing the western territories into sections to allow the sale of land. The PLSS divided states into township grids which were further divided into sections and fractions of sections.

Napoleon Bonaparte founded continental Europe's first cadastre in 1808. This gathered data on the number of parcels of land, their value, land usage, and names. This system soon spread around Europe.

A railroad surveying party at Russel's Tank, Arizona in the 1860s

Robert Torrens introduced the Torrens system in South Australia in 1858. Torrens intended to simplify land transactions and provide reliable titles via a centralized register of land. The Torrens system was adopted in several other nations of the English-speaking world. Surveying became increasingly important with the arrival of railroads in the 1800s. Surveying was necessary so that railroads could plan technologically and financially viable routes.

20th Century

A German engineer surveying during the First World War, 1918

At the beginning of the century surveyors had improved the older chains and ropes, but still faced the problem of accurate measurement of long distances. Dr Trevor Lloyd Wadley developed the Tellurometer during the 1950s. It measures long distances using two microwave transmitter/receivers. During the late 1950s Geodimeter introduced electronic distance measurement (EDM) equipment. EDM units use a multi frequency phase shift of light waves to find a distance. These instruments saved the need for days or weeks of chain measurement by measuring between points kilometers apart in one go.

Advances in electronics allowed miniaturization of EDM. In the 1970s the first instruments combining angle and distance measurement appeared, becoming known as total stations. Manufacturers added more equipment by degrees, bringing improvements in accuracy and speed of measurement. Major advances include tilt compensators, data recorders, and on-board calculation programs.

The first Satellite positioning system was the US Navy TRANSIT system. The first successful launch took place in 1960. The system's main purpose was to provide position information to Polaris missile submarines. Surveyors found they could use field receivers to determine the location of a point. Sparse satellite cover and large equipment made observations laborious, and inaccurate. The main use was establishing benchmarks in remote locations.

The US Air Force launched the first prototype satellites of the Global Positioning System (GPS) in 1978. GPS used a larger constellation of satellites and improved signal transmission to provide more accuracy. Early GPS observations required several hours of observations by a static receiver to reach survey accuracy requirements. Recent improvements to both satellites and receivers allow Real Time Kinematic (RTK) surveying. RTK surveys get high-accuracy measurements by using a fixed base station and a second roving antenna. The position of the roving antenna can be tracked.

21st Century

The theodolite, total station, and RTK GPS survey remain the primary methods in use.

Remote sensing and satellite imagery continue to improve and become cheaper, allowing more commonplace use. Prominent new technologies include three-dimensional (3D) scanning and use of lidar for topographical surveys. UAV technology along with photogrammetric image processing is also appearing.

Hardware

The main surveying instruments in use around the world are the theodolite, measuring tape, total station, 3D scanners, GPS/GNSS, level and rod. Most instruments screw onto a tripod when in use. Tape measures are often used for measurement of smaller distances. 3D scanners and various forms of aerial imagery are also used.

Surveying Equipment. Clockwise from upper left: Optical Theodolite, Robotic total station, RTK GPS Base station, Optical level.

The Theodolite is an instrument for the measurement of angles. It uses two separate *circles*, *protractors* or *alidades* to measure angles in the horizontal and the vertical plane. A telescope mounted on trunnions is aligned vertically with the target object. The whole upper section rotates for horizontal alignment. The vertical circle measures the angle that the telescope makes against the vertical, known as the vertical angle. The horizontal circle uses an upper and lower plate. When beginning the survey, the surveyor points the instrument in a known direction (bearing), and clamps the lower plate in place. The instrument can then rotate to measure the bearing to other objects. If no bearing is known or direct angle measurement is wanted, the instrument can be set to zero during the initial sight. It will then read the angle between the initial object, the theodolite itself, and the item that the telescope aligns with.

The Gyrotheodolite is a form of theodolite that uses a gyroscope to orient itself in the absence of reference marks. It is used in underground applications.

The total station is a development of the theodolite with an electronic distance measurement device (EDM). A total station can be used for leveling when set to the horizontal plane. Since their introduction, total stations have shifted from optical-mechanical to fully electronic devices.

Modern top-of-the-line total stations no longer need a reflector or prism to return the light pulses used for distance measurements. They are fully robotic, and can even e-mail point data to a remote computer and connect to satellite positioning systems, such as Global Positioning System. Real Time Kinematic GPS systems have increased the speed of surveying, but they are still only horizontally accurate to about 20 mm and vertically to 30–40 mm.

GPS surveying differs from other GPS uses in the equipment and methods used. Static GPS uses two receivers placed in position for a considerable length of time. The long span of time lets the receiver compare measurements as the satellites orbit. The changes as the satellites orbit also provide the measurement network with well conditioned geometry. This produces an accurate baseline that can be over 20 km long. RTK surveying uses one static antenna and one roving antenna. The static antenna tracks changes in the satellite positions and atmospheric conditions. The surveyor uses the roving antenna to measure the points needed for the survey. The two antennas use a radio link that allows the static antenna to send corrections to the roving antenna. The roving antenna then applies those corrections to the GPS signals it is receiving to calculate its own position. RTK surveying covers smaller distances than static methods. This is because divergent conditions further away from the base reduce accuracy.

Surveying instruments have characteristics that make them suitable for certain uses. Theodolites and levels are often used by constructors rather than surveyors in first world countries. The constructor can perform simple survey tasks using a relatively cheap instrument. Total stations are workhorses for many professional surveyors because they are versatile and reliable in all conditions. The productivity improvements from a GPS on large scale surveys makes them popular for major infrastructure or data gathering projects. One-person robotic-guided total stations allow surveyors to measure without extra workers to aim the telescope or record data. A fast but expensive way to measure large areas is with a helicopter, using a GPS to record the location of the helicopter and a laser scanner to measure the ground. To increase precision, surveyors place beacons on the ground (about 20 km (12 mi) apart). This method reaches precisions between 5–40 cm (depending on flight height).

Surveyors use ancillary equipment such as: tripods and instrument stands; staves and beacons used for sighting purposes; PPE; vegetation clearing equipment; digging implements for finding survey markers buried over time; hammers for placements of markers in various surfaces and structures; and portable radios for communication over long lines of sight.

Software

Land surveyors, construction professionals and civil engineers using total station, GPS, 3D scanners and other collector data use Land Surveying Software to increase efficiency, accuracy and productivity. Land Surveying Software is a staple of contemporary land surveying.

Surveying Techniques

A standard Brunton Geo compass, still used commonly today by geographers, geologists and surveyors for field-based measurements

Surveyors determine the position of objects by measuring angles and distances. The factors that can affect the accuracy of their observations are also measured. They then use this data to create vectors, bearings, co-ordinates, elevations, areas, volumes, plans and maps. Measurements are often split into horizontal and vertical components to simplify calculation. GPS and astronomic measurements also need measurement of a time component.

Distance Measurement

Example of modern equipment for surveying (Field-Map technology): GPS, laser rangefinder and field computer allows surveying as well as cartography (creation of map in real-time) and field data collection.

Before EDM devices, distances were measured using a variety of means. These included chains having links of a known length such as a Gunter's chain, or measuring tapes made of steel or invar. To measure horizontal distances, these chains or tapes were pulled taut to reduce sagging and slack. The distance had to be adjusted for heat expansion. Attempts to hold the measuring instrument level would also be made. When measuring up a slope, the surveyor might have to "break" (break chain) the measurement- use an increment less than the total length of the chain. Perambulators, or measuring wheels, were used to measure longer distances but not to a high level of accuracy. Tacheometry is the science of measuring distances by measuring the angle between two ends of an object with a known size. It was sometimes used before to the invention of EDM where rough ground made chain measurement impractical.

Angle Measurement

Historically, horizontal angles were measured by using a compass to provide a magnetic bearing. The deflection from the bearing was recorded. Later, more precise scribed discs later improved better angular resolution. Mounting telescopes with reticles atop the disc allowed more precise sighting. Levels and calibrated circles allowed measurement of vertical angles. verniers allowed measurement to a fraction of a degree, such as with a turn-of-the-century transit.

The Plane table provided a graphical method of recording and measuring angles, which reduced the amount of mathematics required. In 1829 Francis Ronalds invented a reflecting instrument for recording angles graphically by modifying the octant.

By observing the bearing from every vertex in a figure, a surveyor can measure around the figure. The final observation will be between the two points first observed, except with a 180° difference. This is called a *close*. If the first and last bearings are different, this shows the error in the survey, called the *angular misclose*. The surveyor can use this information to prove that the work meets the expected standards.

Levelling

Center for Operational Oceanographic Products and Services staff member conducts tide station leveling in support of the US Army Corp of Engineers in Richmond, Maine.

The simplest method for measuring height is with an altimeter using air pressure to find height. When more precise measurements are needed, means like precise levels (also known as differential leveling) are used. When precise leveling, a series of measurements between two points are taken using an instrument and a measuring rod. Differences in height between the measurements are added and subtracted in a series to get the net difference in elevation between the two endpoints. With the Global Positioning System (GPS), elevation can be measured with satellite receivers. Usually GPS is somewhat less accurate than traditional precise leveling, but may be similar over long distances.

When using an optical level, the endpoint may be out of the effective range of the instrument. There may be obstructions or large changes of elevation between the endpoints. In these situations, extra setups are needed. *Turning* is a term used when referring to moving the level to take an elevation shot from a different location. To "turn" the level, one must first take a reading and record the elevation of the point the rod is located on. While the rod is being kept in exactly the same location, the level is moved to a new location where the rod is still visible. A reading is taken from the new location of the level and the height difference is used to find the new elevation of the level gun. This is repeated until the series of measurements is completed. The level must be horizontal to get a valid measurement. Because of this, if the horizontal crosshair of the instrument is lower than the base of the rod, the surveyor will not be able to sight the rod and get a reading. The rod can usually be raised up to 25 feet high, allowing the level to be set much higher than the base of the rod.

Determining Position

The primary way of determining one's position on the earth's surface when no known positions are nearby is by astronomic observations. Observations to the sun, moon and stars could all be made using navigational techniques. Once the instrument's position and bearing to a star is determined, the bearing can be transferred to a reference point on the earth. The point can then be used as a base for further observations. Survey-accurate astronomic positions were difficult to observe and calculate and so tended to be a base off which many other measurements were made. Since the advent of the GPS system, astronomic observations are rare as GPS allows adequate positions to be determined over most of the surface of the earth.

Reference Networks

Few survey positions are derived from first principles. Instead, most surveys points are measured relative to previous measured points. This forms a reference or *control* network where each point can be used by a surveyor to determine their own position when beginning a new survey.

Survey points are usually marked on the earth's surface by objects ranging from small nails driven into the ground to large beacons that can be seen from long distances. The surveyors can set up their instruments on this position and measure to nearby objects. Sometimes a tall, distinctive feature such as a steeple or radio aerial has its position calculated as a reference point that angles can be measured against.

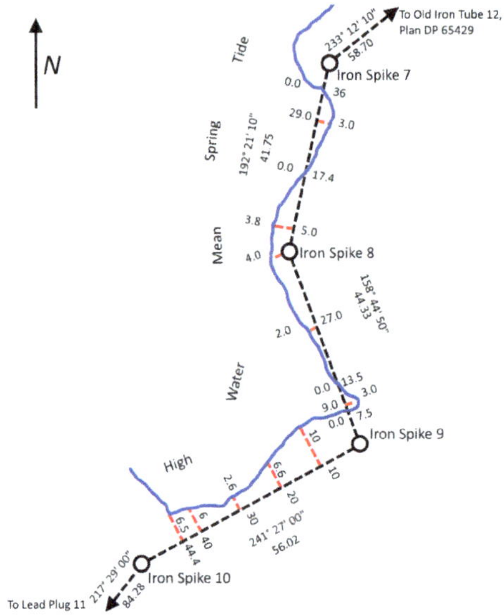

A survey using traverse and offset measurements to record the location of the shoreline shown in blue. Black dashed lines are traverse measurements between reference points (black circles). The red lines are offsets measured at right angles to the traverse lines.

Triangulation is a method of horizontal location favoured in the days before EDM and GPS measurement. It can determine distances, elevations and directions between distant objects. Since the early days of surveying, this was the primary method of determining accurate positions of objects for topographic maps of large areas. A surveyor first needs to know the horizontal distance between two of the objects, known as the baseline. Then the heights, distances and angular position of other objects can be derived, as long as they are visible from one of the original objects. High-accuracy transits or theodolites were used, and angle measurements repeated for increased accuracy.

Offsetting is an alternate method of determining position of objects, and was often used to measure imprecise features such as riverbanks. The surveyor would mark and measure two known positions on the ground roughly parallel to the feature, and mark out a baseline between them. At regular intervals, a distance was measured at right angles from the first line to the feature. The measurements could then be plotted on a plan or map, and the points at the ends of the offset lines could be joined to show the feature.

Traversing is a common method of surveying smaller areas. The surveyor starts from an old reference mark or known position and places a network of reference marks covering the survey area. They then measure bearings and distances between the reference marks, and to the target features. Most traverses form a loop pattern or link between two prior reference marks to allow the surveyor to check their measurements are correct.

Datum and Coordinate Systems

Many surveys do not calculate positions on the surface of the earth, but instead measure the relative positions of objects. However, often the surveyed items need to be compared to outside data, such as boundary lines or previous surveys objects. The oldest way of describing a position is via latitude and longitude, and often a height above sea level. As the surveying profession grew it created Cartesian coordinate systems to simplify the mathematics for surveys over small parts of the earth. The simplest coordinate systems assume that the earth is flat and measure from an arbitrary point, known as a 'datum' (singular form of data). The coordinate system allows easy calculation of the distances and direction between objects over small areas. Large areas distort due to the earth's curvature. North is often defined as true north at the datum.

For larger regions, it is necessary to model the shape of the earth using an ellipsoid or a geoid. Many countries have created coordinate-grids customized to lessen error in their area of the earth.

Errors and Accuracy

A basic tenet of surveying is that no measurement is perfect, and that there will always be a small amount of error. There are three classes of survey errors:

- *Gross errors or blunders:* Errors made by the surveyor during the survey. Upsetting the instrument, misaiming a target, or writing down a wrong measurement are all gross errors. A large gross error may reduce the accuracy to an unacceptable level. Therefore, surveyors use redundant measurements and independent checks to detect these errors early in the survey.
- *Systematic:* Errors that follow a consistent pattern. Examples include effects of temperature on a chain or EDM measurement, or a poorly adjusted spirit-level causing a tilted instrument or target pole. Systematic errors that have known effects can be compensated or corrected.
- *Random:* Random errors are small unavoidable fluctuations. They are caused by imperfections in measuring equipment, eyesight, and conditions. They can be minimized by redundancy of measurement and avoiding unstable conditions. Random errors tend to cancel each other out, but checks must be made to ensure they are not propagating from one measurement to the next.

Surveyors avoid these errors by regular checks on their equipment, using consistent methods, and by good design of their reference network. Redundancy of measurements allows the use of averaging and allows outlier measurements to be discarded. Independent checks like measuring a point from two or more locations or using two different methods are used. Errors can be detected by comparing the results of the two measurements.

Once the surveyor has calculated the level of the errors in his work, it is adjusted. This is the process of distributing the error between all measurements. Each observation is weighted according to how much of the total error it is likely to have caused and part of that error is allocated to it in a proportional way. The most common methods of adjustment are the Bowditch method, also known as the compass rule, and the Principle of least squares method.

The Surveyor must be able to distinguish between accuracy and precision. In the United States, surveyors and civil engineers use units of feet wherein a survey foot breaks down into 10ths and 100ths. Many deed descriptions containing distances are often expressed using these units (125.25 ft). On the subject of accuracy, surveyors are often held to a standard of one one-hundredth of a foot; about 1/8 inch. Calculation and mapping tolerances are much smaller wherein achieving near-perfect closures are desired. Though tolerances will vary from project to project, in the field and day to day usage beyond a 100th of a foot is often impractical.

Types of Surveys

Local professional organisation or regulatory bodies classify specializations of surveying in different ways. Broad groups are:

- *As-built survey*: a survey that documents the location of recently constructed elements of a construction project. Asbuilt surveys are done for record, completion evaluation and payment purposes. An as-built survey is also known as a 'works as executed survey'. As built surveys are often presented in red or redline and laid over existing plans for comparison with design information.

- *Cadastral or boundary surveying*: a survey that establishes or re-establishes boundaries of a parcel using a legal description. It involves the setting or restoration of monuments or markers at the corners or along the lines of the parcel. These take the form of iron rods, pipes, or concrete monuments in the ground, or nails set in concrete or asphalt. The *ALTA/ACSM* Land Title Survey is a standard proposed by the American Land Title Association and the American Congress on Surveying and Mapping. It incorporates elements of the boundary survey, mortgage survey, and topographic survey.

- *Control surveying*: Control surveys establish reference points to use as starting positions for future surveys. Most other forms of surveying will contain elements of control surveying.

- *Construction surveying*

- *Deformation survey:* a survey to determine if a structure or object is changing shape or moving. First the positions of points on an object are found. A period of time is allowed to pass and the positions are then re-measured and calculated. Then a comparison between the two sets of positions is made.

- *Dimensional control survey:* This is a type of survey conducted in or on an non-level surface. Common in the oil and gas industry to replace old or damaged pipes on a like-for-like basis. The advantage of dimensional control survey is that the instrument used to conduct the survey does not need to be level. This is useful in the off-shore industry, as not all platforms are fixed and are thus subject to movement.
- *Engineering surveying*: topographic, layout, and as-built surveys associated with engineering design. They often need geodetic computations beyond normal civil engineering practice.
- *Foundation survey*: a survey done to collect the positional data on a foundation that has been poured and is cured. This is done to ensure that the foundation was constructed in the location, and at the elevation, authorized in the *plot plan*, *site plan*, or *subdivision plan*.
- *Hydrographic survey*: a survey conducted with the purpose of mapping the shoreline and bed of a body of water. Used for navigation, engineering, or resource management purposes.
- *Leveling*: either finds the elevation of a given point or establish a point at a given elevation.
- *LOMA survey*: Survey to change base flood line, removing property from a SFHA special flood hazard area.
- *Measured survey* : a building survey to produce plans of the building. such a survey may be conducted before renovation works, for commercial purpose, or at end of the construction process.
- *Mining surveying*: Mining surveying includes directing the digging of mine shafts and galleries and the calculation of volume of rock. It uses specialised techniques due to the restraints to survey geometry such as vertical shafts and narrow passages.
- *Mortgage survey:* A *mortgage survey* or *physical survey* is a simple survey that delineates land boundaries and building locations. It checks for *encroachment*, building setback restrictions and shows nearby flood zones. In many places a mortgage survey is a precondition for a mortgage loan.
- *Photographic control survey*: A survey that creates reference marks visible from the air to allow aerial photographs to be rectified.
- *Stakeout, Layout or Setout*: an element of many other surveys where the calculated or proposed position of an object is marked on the ground. This can be temporary or permanent. This is an important component of engineering and cadastral surveying.

- *Structural survey*: a detailed inspection to report upon the physical condition and structural stability of a building or structure. It highlights any work needed to maintain it in good repair.
- *Subdivision*: A boundary survey that splits a property into two or more smaller properties.
- *Topographic survey*: a survey that measures the elevation of points on a particular piece of land, and presents them as contour lines on a plot.

Plane and Geodetic Surveying

Based on the considerations and true shape of the earth, surveying is broadly classified into two types.

Plane surveying assumes the earth is flat. Curvature and spheroidal shape of the earth is neglected. In this type of surveying all triangles formed by joining survey lines are considered as plane triangles. It is employed for small survey works where errors due to the earth's shape are too small to matter.

In *geodetic surveying* the curvature of the earth is taken into account while calculating reduced levels, angles, bearings and distances. This type of surveying is usually employed for large survey works. Survey works up to 100 square miles (260 square kilometers) are treated as plane and beyond that are treated as geodetic. In geodetic surveying necessary corrections are applied to reduced levels, bearings and other observations.

The Surveying Profession

An all-female surveying crew in Idaho, 1918

The basic principles of surveying have changed little over the ages, but the tools used by surveyors have evolved. Engineering, especially civil engineering, often needs surveyors.

Surveyors help determine the placement of roads, railways, reservoirs, dams, pipelines, retaining walls, bridges, and buildings. They establish the boundaries of legal descriptions and political divisions. They also provide advice and data for *geographical information systems* (GIS) that record land features and boundaries.

Surveyors must have a thorough knowledge of algebra, basic calculus, geometry, and trigonometry. They must also know the laws that deal with surveys, real property, and contracts.

Most jurisdictions recognize three different levels of qualification:

Survey assistants or *chainmen* are usually unskilled workers who help the surveyor. They place target reflectors, find old reference marks, and mark points on the ground. The term 'chainman' derives from past use of measuring chains. An assistant would move the far end of the chain under the surveyor's direction.

Survey technicians often operate survey instruments, run surveys in the field, do survey calculations, or draft plans. A technician usually has no legal authority and cannot certify his work. Not all technicians are qualified, but qualifications at the certificate or diploma level are available.

Licensed, registered, or chartered surveyors usually hold a degree or higher qualification. They are often required to pass further exams to join a professional association or to gain certifying status. Surveyors are responsible for planning and management of surveys. They have to ensure that their surveys, or surveys performed under their supervision, meet the legal standards. Many principals of surveying firms hold this status.

Informal Surveying

Not all surveys are carried out by professional surveyors. Depending on the jurisdiction and circumstances, the builders of a structure may set it out themselves. Surveyors often set out the most significant corners of a building. The builders then lay out the rest of the building themselves using simple surveying techniques.

Licensing

Surveying students with their professor at the Helsinki University of Technology in the late 19th century

Licensing requirements vary with jurisdiction, and are commonly consistent within national borders. Prospective surveyors usually have to receive a degree in surveying, followed by a detailed examination of their knowledge of surveying law and principles specific to the region they wish to practice in, and undergo a period of on-the-job training or portfolio building before they are awarded a license to practise. Licensed surveyors usually receive a post nominal, which varies depending on where they qualified. The system has replaced older apprenticeship systems.

A licensed land surveyor is generally required to sign and seal all plans. The state dictates the format, showing their name and registration number.

In many jurisdictions, surveyors must mark their registration number on survey monuments when setting boundary corners. Monuments take the form of capped iron rods, concrete monuments, or nails with washers.

Surveying Institutions

Most countries' governments regulate at least some forms of surveying. Their survey agencies establish regulations and standards. Standards control accuracy, surveying credentials, monumentation of boundaries and maintenance of geodetic networks. Many nations devolve this authority to regional entities or states/provinces. Cadastral surveys tend to be the most regulated because of the permanence of the work. Lot boundaries established by cadastral surveys may stand for hundreds of years without modification.

Most jurisdictions also have a form of professional institution representing local surveyors. These institutes often endorse or license potential surveyors, as well as set and enforce ethical standards. The largest institution is the International Federation of Surveyors (Abbreviated FIG, for French: *Fédération Internationale des Géomètres*). They represent the survey industry worldwide.

Building Surveying

Most English-speaking countries consider building surveying a distinct profession. They have their own professional associations and licensing requirements. Building surveyors focus on investigating the condition of buildings as well as legal compliance work.

Cadastral Surveying

One of the primary roles of the land surveyor is to determine the boundary of real property on the ground. The surveyor must determine where the adjoining landowners wish to put the boundary. The boundary is established in legal documents and plans prepared by attorneys, engineers, and land surveyors. The surveyor then puts monuments on the corners of the new boundary. They might also find or resurvey the corners of the property monumented by prior surveys.

Cadastral land surveyors are licensed by governments. The cadastral survey branch of the Bureau of Land Management (BLM) conducts most cadastral surveys in the United States. They consult with Forest Service, National Park Service, Army Corps of Engineers, Bureau of Indian Affairs, Fish and Wildlife Service, Bureau of Reclamation, and others. The BLM used to be known as the General Land Office (GLO).

In states organized per the Public Land Survey System (PLSS), surveyors must carry out BLM cadastral surveys under that system.

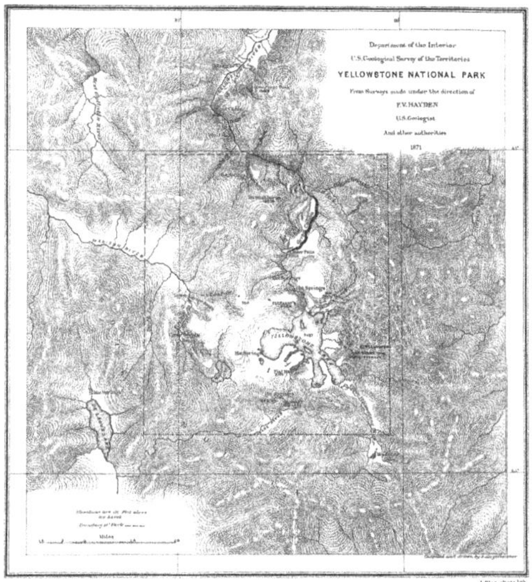

F.V. Hayden's map of Yellowstone National Park, 1871. His surveys were a significant basis for establishing the park in 1872.

Cadastral surveyors often have to work around changes to the earth that obliterate or damage boundary monuments. When this happens, they must consider evidence that is not recorded on the title deed. This is known as extrinsic evidence.

Remote Sensing

Remote sensing is the acquisition of information about an object or phenomenon without making physical contact with the object and thus in contrast to on site observation. Remote sensing is used in numerous fields, including geography and most Earth Science disciplines (for example, hydrology, ecology, oceanography, glaciology, geology); it also has military, intelligence, commercial, economic, planning, and humanitarian applications. In modern usage, the term generally refers to the use of aerial sensor technologies to detect and classify objects on Earth (both on the surface, and in the atmosphere and oceans) by means of propagated signals (e.g. electromagnetic radiation). It may be split into active remote sensing (when a signal is first emitted from aircraft or satellites) or passive (e.g. sunlight) when information is merely recorded.

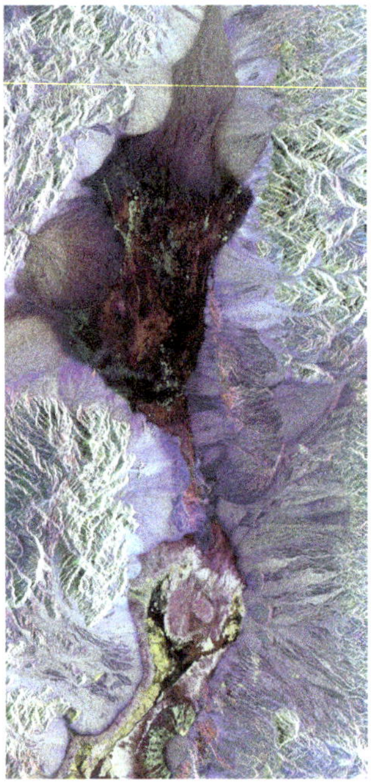

Synthetic aperture radar image of Death Valley colored using polarimetry.

Overview

Passive sensors gather radiation that is emitted or reflected by the object or surrounding areas. Reflected sunlight is the most common source of radiation measured by passive sensors. Examples of passive remote sensors include film photography, infrared, charge-coupled devices, and radiometers. Active collection, on the other hand, emits energy in order to scan objects and areas whereupon a sensor then detects and measures the radiation that is reflected or backscattered from the target. RADAR and LiDAR are examples of active remote sensing where the time delay between emission and return is measured, establishing the location, speed and direction of an object.

This video is about how Landsat was used to identify areas of conservation in the Democratic Republic of the Congo, and how it was used to help map an area called MLW in the north

Remote sensing makes it possible to collect data of dangerous or inaccessible areas. Remote sensing applications include monitoring deforestation in areas such as the Amazon Basin, glacial features in Arctic and Antarctic regions, and depth sounding of coastal and ocean depths. Military collection during the Cold War made use of stand-off collection of data about dangerous border areas. Remote sensing also replaces costly and slow data collection on the ground, ensuring in the process that areas or objects are not disturbed.

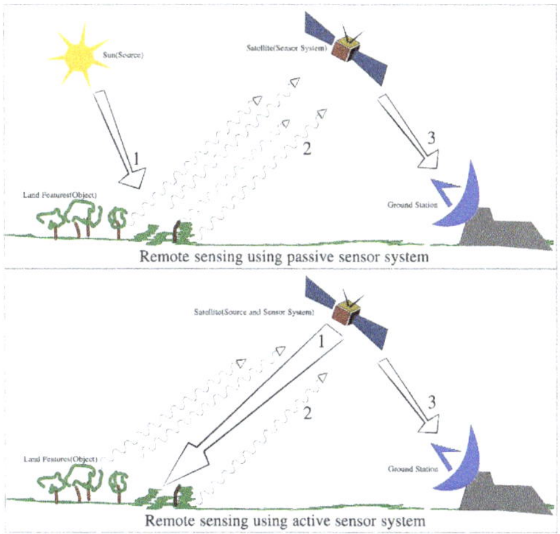

Illustration of Remote Sensing

Orbital platforms collect and transmit data from different parts of the electromagnetic spectrum, which in conjunction with larger scale aerial or ground-based sensing and analysis, provides researchers with enough information to monitor trends such as El Niño and other natural long and short term phenomena. Other uses include different areas of the earth sciences such as natural resource management, agricultural fields such as land usage and conservation, and national security and overhead, ground-based and stand-off collection on border areas.

Data Acquisition Techniques

The basis for multispectral collection and analysis is that of examined areas or objects that reflect or emit radiation that stand out from surrounding areas.

Applications of Remote Sensing

- Conventional radar is mostly associated with aerial traffic control, early warning, and certain large scale meteorological data. Doppler radar is used by local law enforcements' monitoring of speed limits and in enhanced meteorological collection such as wind speed and direction within weather systems in addition to precipitation location and intensity. Other types of active collection includes plasmas in the ionosphere. Interferometric synthetic aperture radar is used to produce precise digital elevation models of large scale terrain.

- Laser and radar altimeters on satellites have provided a wide range of data.

By measuring the bulges of water caused by gravity, they map features on the seafloor to a resolution of a mile or so. By measuring the height and wavelength of ocean waves, the altimeters measure wind speeds and direction, and surface ocean currents and directions.

- Ultrasound (acoustic) and radar tide gauges measure sea level, tides and wave direction in coastal and offshore tide gauges.
- Light detection and ranging (LIDAR) is well known in examples of weapon ranging, laser illuminated homing of projectiles. LIDAR is used to detect and measure the concentration of various chemicals in the atmosphere, while airborne LIDAR can be used to measure heights of objects and features on the ground more accurately than with radar technology. Vegetation remote sensing is a principal application of LIDAR.
- Radiometers and photometers are the most common instrument in use, collecting reflected and emitted radiation in a wide range of frequencies. The most common are visible and infrared sensors, followed by microwave, gamma ray and rarely, ultraviolet. They may also be used to detect the emission spectra of various chemicals, providing data on chemical concentrations in the atmosphere.
- Stereographic pairs of aerial photographs have often been used to make topographic maps by imagery and terrain analysts in trafficability and highway departments for potential routes, in addition to modelling terrestrial habitat features.
- Simultaneous multi-spectral platforms such as Landsat have been in use since the 70's. These thematic mappers take images in multiple wavelengths of electro-magnetic radiation (multi-spectral) and are usually found on Earth observation satellites, including (for example) the Landsat program or the IKONOS satellite. Maps of land cover and land use from thematic mapping can be used to prospect for minerals, detect or monitor land usage, detect invasive vegetation, deforestation, and examine the health of indigenous plants and crops, including entire farming regions or forests. Landsat images are used by regulatory agencies such as KYDOW to indicate water quality parameters including Secchi depth, chlorophyll a density and total phosphorus content. Weather satellites are used in meteorology and climatology.
- Hyperspectral imaging produces an image where each pixel has full spectral information with imaging narrow spectral bands over a contiguous spectral range. Hyperspectral imagers are used in various applications including mineralogy, biology, defence, and environmental measurements.
- Within the scope of the combat against desertification, remote sensing allows to follow-up and monitor risk areas in the long term, to determine desertification factors, to support decision-makers in defining relevant measures of environmental management, and to assess their impacts.

Geodetic

- Overhead geodetic collection was first used in aerial submarine detection and gravitational data used in military maps. This data revealed minute perturbations in the Earth's gravitational field (geodesy) that may be used to determine changes in the mass distribution of the Earth, which in turn may be used for geological studies.

Acoustic and Near-acoustic

- Sonar: *passive sonar*, listening for the sound made by another object (a vessel, a whale etc.); *active sonar*, emitting pulses of sounds and listening for echoes, used for detecting, ranging and measurements of underwater objects and terrain.
- Seismograms taken at different locations can locate and measure earthquakes (after they occur) by comparing the relative intensity and precise timings.
- Ultrasound: Ultrasound sensors, that emit high frequency pulses and listening for echoes, used for detecting water waves and water level, as in tide gauges or for towing tanks.

To coordinate a series of large-scale observations, most sensing systems depend on the following: platform location and the orientation of the sensor. High-end instruments now often use positional information from satellite navigation systems. The rotation and orientation is often provided within a degree or two with electronic compasses. Compasses can measure not just azimuth (i. e. degrees to magnetic north), but also altitude (degrees above the horizon), since the magnetic field curves into the Earth at different angles at different latitudes. More exact orientations require gyroscopic-aided orientation, periodically realigned by different methods including navigation from stars or known benchmarks.

Data Processing

Generally speaking, remote sensing works on the principle of the inverse problem. While the object or phenomenon of interest (the state) may not be directly measured, there exists some other variable that can be detected and measured (the observation) which may be related to the object of interest through a calculation. The common analogy given to describe this is trying to determine the type of animal from its footprints. For example, while it is impossible to directly measure temperatures in the upper atmosphere, it is possible to measure the spectral emissions from a known chemical species (such as carbon dioxide) in that region. The frequency of the emissions may then be related via thermodynamics to the temperature in that region.

The quality of remote sensing data consists of its spatial, spectral, radiometric and temporal resolutions.

Spatial Resolution

The size of a pixel that is recorded in a raster image – typically pixels may correspond to square areas ranging in side length from 1 to 1,000 metres (3.3 to 3,280.8 ft).

Spectral Resolution

The wavelength width of the different frequency bands recorded – usually, this is related to the number of frequency bands recorded by the platform. Current Landsat collection is that of seven bands, including several in thc infra-red spectrum, ranging from a spectral resolution of 0.07 to 2.1 μm. The Hyperion sensor on Earth Observing-1 resolves 220 bands from 0.4 to 2.5 μm, with a spectral resolution of 0.10 to 0.11 μm per band.

Radiometric Resolution

The number of different intensities of radiation the sensor is able to distinguish. Typically, this ranges from 8 to 14 bits, corresponding to 256 levels of the gray scale and up to 16,384 intensities or "shades" of colour, in each band. It also depends on the instrument noise.

Temporal Resolution

The frequency of flyovers by the satellite or plane, and is only relevant in time-series studies or those requiring an averaged or mosaic image as in deforesting monitoring. This was first used by the intelligence community where repeated coverage revealed changes in infrastructure, the deployment of units or the modification/introduction of equipment. Cloud cover over a given area or object makes it necessary to repeat the collection of said location.

In order to create sensor-based maps, most remote sensing systems expect to extrapolate sensor data in relation to a reference point including distances between known points on the ground. This depends on the type of sensor used. For example, in conventional photographs, distances are accurate in the center of the image, with the distortion of measurements increasing the farther you get from the center. Another factor is that of the platen against which the film is pressed can cause severe errors when photographs are used to measure ground distances. The step in which this problem is resolved is called georeferencing, and involves computer-aided matching of points in the image (typically 30 or more points per image) which is extrapolated with the use of an established benchmark, "warping" the image to produce accurate spatial data. As of the early 1990s, most satellite images are sold fully georeferenced.

In addition, images may need to be radiometrically and atmospherically corrected.

Radiometric Correction

Allows to avoid radiometric errors and distortions. The illumination of objects on the Earth surface is uneven because of different properties of the relief. This factor is taken into account in the method of radiometric distortion correction. Radiometric correction gives a scale to the pixel values, e. g. the monochromatic scale of 0 to 255 will be converted to actual radiance values.

Topographic Correction (also called terrain correction)

In rugged mountains, as a result of terrain, the effective illumination of pixels varies considerably. In a remote sensing image, the pixel on the shady slope receives weak illumination and has a low radiance value, in contrast, the pixel on the sunny slope receives strong illumination and has a high radiance value. For the same object, the pixel radiance value on the shady slope will be different from that on the sunny slope. Additionally, different objects may have similar radiance values. These ambiguities seriously affected remote sensing image information extraction accuracy in mountainous areas. It became the main obstacle to further application of remote sensing images. The purpose of topographic correction is to eliminate this effect, recovering the true reflectivity or radiance of objects in horizontal conditions. It is the premise of quantitative remote sensing application.

Atmospheric Correction

Elimination of atmospheric haze by rescaling each frequency band so that its minimum value (usually realised in water bodies) corresponds to a pixel value of 0. The digitizing of data also makes it possible to manipulate the data by changing gray-scale values.

Interpretation is the critical process of making sense of the data. The first application was that of aerial photographic collection which used the following process; spatial measurement through the use of a light table in both conventional single or stereographic coverage, added skills such as the use of photogrammetry, the use of photomosaics, repeat coverage, Making use of objects' known dimensions in order to detect modifications. Image Analysis is the recently developed automated computer-aided application which is in increasing use.

Object-Based Image Analysis (OBIA) is a sub-discipline of GIScience devoted to partitioning remote sensing (RS) imagery into meaningful image-objects, and assessing their characteristics through spatial, spectral and temporal scale.

Old data from remote sensing is often valuable because it may provide the only long-term data for a large extent of geography. At the same time, the data is often complex to interpret, and bulky to store. Modern systems tend to store the data digitally, often with lossless compression. The difficulty with this approach is that the data is fragile, the format may be archaic, and the data may be easy to falsify. One of the best systems for archiving data series is as computer-generated machine-readable ultrafiche, usually in typefonts such as OCR-B, or as digitized half-tone images. Ultrafiches survive well in standard libraries, with lifetimes of several centuries. They can be created, copied, filed and retrieved by automated systems. They are about as compact as archival magnetic media, and yet can be read by human beings with minimal, standardized equipment.

Data Processing Levels

To facilitate the discussion of data processing in practice, several processing "levels" were first defined in 1986 by NASA as part of its Earth Observing System and steadily adopted since then, both internally at NASA (e. g.,) and elsewhere (e. g.,); these definitions are:

Level	Description
0	Reconstructed, unprocessed instrument and payload data at full resolution, with any and all communications artifacts (e. g., synchronization frames, communications headers, duplicate data) removed.
1a	Reconstructed, unprocessed instrument data at full resolution, time-referenced, and annotated with ancillary information, including radiometric and geometric calibration coefficients and georeferencing parameters (e. g., platform ephemeris) computed and appended but not applied to the Level 0 data (or if applied, in a manner that level 0 is fully recoverable from level 1a data).
1b	Level 1a data that have been processed to sensor units (e. g., radar backscatter cross section, brightness temperature, etc.); not all instruments have Level 1b data; level 0 data is not recoverable from level 1b data.
2	Derived geophysical variables (e. g., ocean wave height, soil moisture, ice concentration) at the same resolution and location as Level 1 source data.
3	Variables mapped on uniform spacetime grid scales, usually with some completeness and consistency (e. g., missing points interpolated, complete regions mosaicked together from multiple orbits, etc.).
4	Model output or results from analyses of lower level data (i. e., variables that were not measured by the instruments but instead are derived from these measurements).

A Level 1 data record is the most fundamental (i. e., highest reversible level) data record that has significant scientific utility, and is the foundation upon which all subsequent data sets are produced. Level 2 is the first level that is directly usable for most scientific applications; its value is much greater than the lower levels. Level 2 data sets tend to be less voluminous than Level 1 data because they have been reduced temporally, spatially, or spectrally. Level 3 data sets are generally smaller than lower level data sets and thus can be dealt with without incurring a great deal of data handling overhead. These data tend to be generally more useful for many applications. The regular spatial and temporal organization of Level 3 datasets makes it feasible to readily combine data from different sources.

While these processing levels are particularly suitable for typical satellite data processing pipelines, other data level vocabularies have been defined and may be appropriate for more heterogeneous workflows.

History

The modern discipline of remote sensing arose with the development of flight. The balloonist G. Tournachon (alias Nadar) made photographs of Paris from his balloon in 1858. Messenger pigeons, kites, rockets and unmanned balloons were also used for early images. With the exception of balloons, these first, individual images were not particularly useful for map making or for scientific purposes.

The TR-1 reconnaissance/surveillance aircraft.

The *2001 Mars Odyssey* used spectrometers and imagers to hunt for evidence of past or present water and volcanic activity on Mars.

Systematic aerial photography was developed for military surveillance and reconnaissance purposes beginning in World War I and reaching a climax during the Cold War with the use of modified combat aircraft such as the P-51, P-38, RB-66 and the F-4C, or specifically designed collection platforms such as the U2/TR-1, SR-71, A-5 and the OV-1 series both in overhead and stand-off collection. A more recent development is that of increasingly smaller sensor pods such as those used by law enforcement and the military, in both manned and unmanned platforms. The advantage of this approach is that this requires minimal modification to a given airframe. Later imaging technologies would include Infra-red, conventional, Doppler and synthetic aperture radar.

The development of artificial satellites in the latter half of the 20th century allowed remote sensing to progress to a global scale as of the end of the Cold War. Instrumentation aboard various Earth observing and weather satellites such as Landsat, the Nimbus and more recent missions such as RADARSAT and UARS provided global measurements of various data for civil, research, and military purposes. Space probes to other planets have also provided the opportunity to conduct remote sensing studies in extraterrestrial environments, synthetic aperture radar aboard the Magellan spacecraft provided detailed topographic maps of Venus, while instruments aboard SOHO allowed studies to be performed on the Sun and the solar wind, just to name a few examples.

Recent developments include, beginning in the 1960s and 1970s with the development of image processing of satellite imagery. Several research groups in Silicon Valley including NASA Ames Research Center, GTE, and ESL Inc. developed Fourier transform techniques leading to the first notable enhancement of imagery data. In 1999 the first commercial satellite (IKONOS) collecting very high resolution imagery was launched.

Training and Education

At most universities remote sensing is associated with Geography departments. Remote Sensing has a growing relevance in the modern information society. It represents a key technology as part of the aerospace industry and bears increasing economic relevance – new sensors e.g. TerraSAR-X and RapidEye are developed constantly and the demand for skilled labour is increasing steadily. Furthermore, remote sensing exceedingly influences everyday life, ranging from weather forecasts to reports on climate change or natural disasters. As an example, 80% of the German students use the services of Google Earth; in 2006 alone the software was downloaded 100 million times. But studies have shown that only a fraction of them know more about the data they are working with. There exists a huge knowledge gap between the application and the understanding of satellite images. Remote sensing only plays a tangential role in schools, regardless of the political claims to strengthen the support for teaching on the subject. A lot of the computer software explicitly developed for school lessons has not yet been implemented due to its complexity. Thereby, the subject is either not at all integrated into the curriculum or does not pass the step of an interpretation of analogue images. In fact, the subject of remote sensing requires a consolidation of physics and mathematics as well as competences in the fields of media and methods apart from the mere visual interpretation of satellite images.

Many teachers have great interest in the subject "remote sensing", being motivated to integrate this topic into teaching, provided that the curriculum is considered. In many cases, this encouragement fails because of confusing information. In order to integrate remote sensing in a sustainable manner organizations like the EGU or digital earth encourages the development of learning modules and learning portals (e.g. FIS – Remote Sensing in School Lessons or Landmap – Spatial Discovery) promoting media and method qualifications as well as independent working.

Remote Sensing Software

Remote sensing data are processed and analyzed with computer software, known as a remote sensing application. A large number of proprietary and open source applications exist to process remote sensing data. Remote sensing software packages include:

- PCI Geomatica made by PCI Geomatics,
- TacitView from 2d3
- Socet GXP from BAE Systems,
- TNTmips from MicroImages,
- IDRISI from Clark Labs,
- eCognition from Trimble,
- and RemoteView made by Overwatch Textron Systems.
- Dragon/ips is one of the oldest remote sensing packages still available, and is in some cases free.
- ERDAS IMAGINE from Hexagon Geospatial (Separated from Intergraph SG&I),
- ENVI/IDL from Exelis Visual Information Solutions,

Open source remote sensing software includes:

- Opticks (software),
- Orfeo toolbox
- Others mixing remote sensing and GIS capabilities are: GRASS GIS, ILWIS, QGIS, and TerraLook.

According to an NOAA Sponsored Research by Global Marketing Insights, Inc. the most used applications among Asian academic groups involved in remote sensing are as follows: ERDAS 36% (ERDAS IMAGINE 25% & ERMapper 11%); ESRI 30%; ITT Visual Information Solutions ENVI 17%; MapInfo 17%.

Among Western Academic respondents as follows: ESRI 39%, ERDAS IMAGINE 27%, MapInfo 9%, and AutoDesk 7%.

Satellite Imagery

Satellite imagery consists of images of Earth or other planets collected by satellites. Imaging satellites are operated by governments and businesses around the world. Satellite imaging companies sell images under licence. Images are licensed to governments and businesses such as Apple Maps and Google Maps.

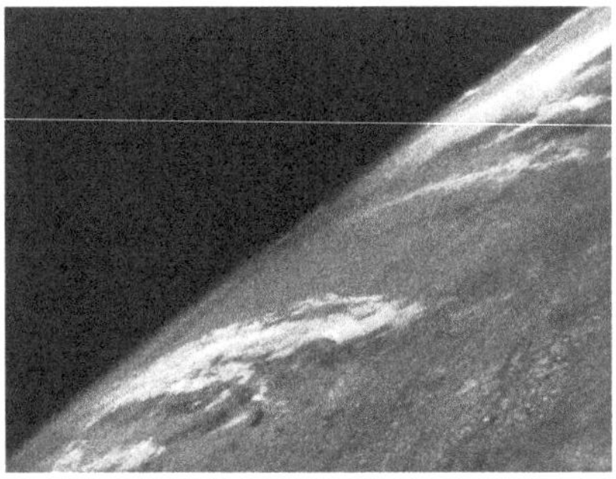

The first images from space were taken on the sub-orbital V-2 rocket flight launched by the U.S. on October 24, 1946.

History

The satellite images were made from pixels. The first crude image taken by the satellite Explorer 6 shows a sunlit area of the Central Pacific Ocean and its cloud cover. The photo was taken when the satellite was about 17,000 mi (27,000 km) above the surface of the earth on August 14, 1959. At the time, the satellite was crossing Mexico.

The first images from space were taken on sub-orbital flights. The U.S-launched V-2 flight on October 24, 1946 took one image every 1.5 seconds. With an apogee of 65 miles (105 km), these photos were from five times higher than the previous record, the 13.7 miles (22 km) by the Explorer II balloon mission in 1935. The first satellite (orbital) photographs of Earth were made on August 14, 1959 by the U.S. Explorer 6. The first satellite photographs of the Moon might have been made on October 6, 1959 by the Soviet satellite Luna 3, on a mission to photograph the far side of the Moon. The Blue Marble photograph was taken from space in 1972, and has become very popular in the media and among the public. Also in 1972 the United States started the Landsat program, the largest program for acquisition of imagery of Earth from space. Landsat Data Continuity Mission, the most recent Landsat satellite, was launched on 11 February 2013. In 1977, the first real time satellite imagery was acquired by the United States's KH-11 satellite system.

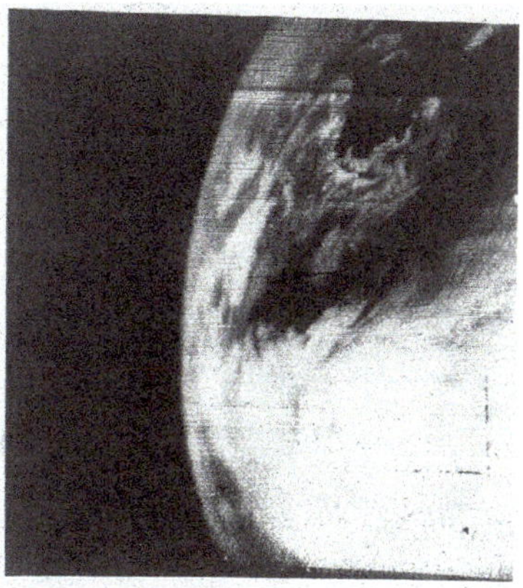

The first television image of Earth from space transmitted by the TIROS-1 weather satellite in 1960.

All satellite images produced by NASA are published by NASA Earth Observatory and are freely available to the public. Several other countries have satellite imaging programs, and a collaborative European effort launched the ERS and Envisat satellites carrying various sensors. There are also private companies that provide commercial satellite imagery. In the early 21st century satellite imagery became widely available when affordable, easy to use software with access to satellite imagery databases was offered by several companies and organizations.

Uses

Satellite photography can be used to produce composite images of an entire hemisphere

...or to map a small area of the Earth, such as this photo of the countryside of Haskell County, Kansas, United States.

Satellite images have many applications in meteorology, oceanography, fishing, agriculture, biodiversity conservation, forestry, landscape, geology, cartography, regional planning, education, intelligence and warfare. Images can be in visible colours and in other spectra. There are also elevation maps, usually made by radar images. Interpretation and analysis of satellite imagery is conducted using specialized remote sensing applications.

Resolution and Data

There are four types of resolution when discussing satellite imagery in remote sensing: spatial, spectral, temporal, and radiometric. Campbell (2002) defines these as follows:

- spatial resolution is defined as the pixel size of an image representing the size of the surface area (i.e. m^2) being measured on the ground, determined by the sensors' instantaneous field of view (IFOV);
- spectral resolution is defined by the wavelength interval size (discreet segment of the Electromagnetic Spectrum) and number of intervals that the sensor is measuring;
- temporal resolution is defined by the amount of time (e.g. days) that passes between imagery collection periods for a given surface location
- Radiometric resolution is defined as the ability of an imaging system to record many levels of brightness (contrast for example) and to the effective bit-depth of the sensor (number of grayscale levels) and is typically expressed as 8-bit (0-255), 11-bit (0-2047), 12-bit (0-4095) or 16-bit (0-65,535).
- Geometric resolution refers to the satellite sensor's ability to effectively image a portion of the Earth's surface in a single pixel and is typically expressed in terms of Ground sample distance, or GSD. GSD is a term containing the overall optical and systemic noise sources and is useful for comparing how well one sensor can "see" an object on the ground within a single pixel. For example, the GSD of Landsat is ~30m, which means the smallest unit that maps to a single pixel within an image is ~30m x 30m. The latest commercial satellite (GeoEye 1) has a GSD of 0.41 m. This compares to a 0.3 m resolution obtained by some early military film based Reconnaissance satellite such as Corona.

The resolution of satellite images varies depending on the instrument used and the altitude of the satellite's orbit. For example, the Landsat archive offers repeated imagery at 30 meter resolution for the planet, but most of it has not been processed from the raw data. Landsat 7 has an average return period of 16 days. For many smaller areas, images with resolution as high as 41 cm can be available.

Satellite imagery is sometimes supplemented with aerial photography, which has higher resolution, but is more expensive per square meter. Satellite imagery can be combined with vector or raster data in a GIS provided that the imagery has been spatially rectified so that it will properly align with other data sets.

Imaging Satellites

GeoEye

GeoEye's GeoEye-1 satellite was launched September 6, 2008. The GeoEye-1 satellite has the high resolution imaging system and is able to collect images with a ground resolution of 0.41 meters (16 inches) in the panchromatic or black and white mode. It collects multispectral or color imagery at 1.65-meter resolution or about 64 inches.

DigitalGlobe

DigitalGlobe's WorldView-2 satellite provides high resolution commercial satellite imagery with 0.46 m spatial resolution (panchromatic only). The 0.46 meters resolution of WorldView-2's panchromatic images allows the satellite to distinguish between objects on the ground that are at least 46 cm apart. Similarly DigitalGlobe's QuickBird satellite provides 0.6 meter resolution (at NADIR) panchromatic images.

DigitalGlobe's WorldView-3 satellite provides high resolution commercial satellite imagery with 0.31 m spatial resolution. WVIII also carries a short wave infrared sensor and an atmospheric sensor

Spot Image

SPOT image of Bratislava

The 3 SPOT satellites in orbit (Spot 2, 4 and 5) provide images with a large choice of resolutions – from 2.5 m to 1 km. Spot Image also distributes multiresolution data from other optical satellites, in particular from Formosat-2 (Taiwan) and Kompsat-2 (South Korea) and from radar satellites (TerraSar-X, ERS, Envisat, Radarsat). Spot Image will also be the exclusive distributor of data from the forthcoming very-high resolution Pleiades satellites with a resolution of 0.50 meter or about 20 inches. The first launch is planned for the end of 2011. The company also offers infrastructures for receiving and processing, as well as added value options.

ASTER

The Advanced Spaceborne Thermal Emission and Reflection Radiometer (ASTER) is an imaging instrument onboard Terra, the flagship satellite of NASA's Earth Observing System (EOS) launched in December 1999. ASTER is a cooperative effort between NASA, Japan's Ministry of Economy, Trade and Industry (METI), and Japan Space Systems (J-spacesystems). ASTER data is used to create detailed maps of land surface temperature, reflectance, and elevation. The coordinated system of EOS satellites, including Terra, is a major component of NASA's Science Mission Directorate and the Earth Science Division. The goal of NASA Earth Science is to develop a scientific understanding of the Earth as an integrated system, its response to change, and to better predict variability and trends in climate, weather, and natural hazards.

- Land surface climatology—investigation of land surface parameters, surface temperature, etc., to understand land-surface interaction and energy and moisture fluxes
- Vegetation and ecosystem dynamics—investigations of vegetation and soil distribution and their changes to estimate biological productivity, understand land-atmosphere interactions, and detect ecosystem change
- Volcano monitoring—monitoring of eruptions and precursor events, such as gas emissions, eruption plumes, development of lava lakes, eruptive history and eruptive potential
- Hazard monitoring—observation of the extent and effects of wildfires, flooding, coastal erosion, earthquake damage, and tsunami damage
- Hydrology—understanding global energy and hydrologic processes and their relationship to global change; included is evapotranspiration from plants
- Geology and soils—the detailed composition and geomorphologic mapping of surface soils and bedrocks to study land surface processes and earth's history
- Land surface and land cover change—monitoring desertification, deforestation, and urbanization; providing data for conservation managers to monitor protected areas, national parks, and wilderness areas

BlackBridge

BlackBridge, previously known as RapidEye, operates a constellation of five satellites, launched in August 2008, the RapidEye constillation contains identical multispectral sensors which are equally calibrated. Therefore, an image from one satellite will be equivalent to an image from any of the other four, allowing for a large amount of imagery to be collected (4 million km^2 per day), and daily revisit to an area. Each travel on the same orbital plane at 630 km, and deliver images in 5 meter pixel size. RapidEye satellite imagery is especially suited for agricultural, environmental, cartographic and disaster management applications. The company not only offers their imagery, but consults with their customers to create services and solutions based on analysis of this imagery .

ImageSat International

Earth Resource Observation Satellites, better known as "EROS" satellites, are lightweight, low earth orbiting, high-resolution satellites designed for fast maneuvering between imaging targets. In the commercial high-resolution satellite market, EROS is the smallest very high resolution satellite; it is very agile and thus enables very high performances. The satellites are deployed in a circular sun-synchronous near polar orbit at an altitude of 510 km (+/- 40 km). EROS satellites imagery applications are primarily for intelligence, homeland security and national development purposes but also employed in a wide range of civilian applications, including: mapping, border control, infrastructure planning, agricultural monitoring, environmental monitoring, disaster response, training and simulations, etc.

EROS A – a high resolution satellite with 1.9-1.2m resolution panchromatic was launched on December 5, 2000.

EROS B - the second generation of Very High Resolution satellites with 70 cm resolution panchromatic, was launched on April 25, 2006.

Meteosat

Model of a first generation Meteosat geostationary satellite.

The Meteosat-2 geostationary weather satellite began operationally to supply imager data on 16 August 1981. Eumetsat has operated the Meteosats since 1987.

- The *Meteosat visible and infrared imager (MVIRI)*, three-channel imager: visible, infrared and water vapour; It operates on the first generation Meteosat, Meteosat-7 being still active.
- The 12-channel *Spinning Enhanced Visible and Infrared Imager (SEVIRI)* includes similar channels to those used by MVIRI, providing continuity in climate data over three decades; Meteosat Second Generation (MSG).

- The *Flexible Combined Imager (FCI)* on Meteosat Third Generation (MTG) will also include similar channels, meaning that all three generations will have provided over 60 years of climate data.

Disadvantages

Because the total area of the land on Earth is so large and because resolution is relatively high, satellite databases are huge and image processing (creating useful images from the raw data) is time-consuming. Depending on the sensor used, weather conditions can affect image quality: for example, it is difficult to obtain images for areas of frequent cloud cover such as mountain-tops. For such reasons, publicly available satellite image datasets are typically processed for visual or scientific commercial use by third parties.

Commercial satellite companies do not place their imagery into the public domain and do not sell their imagery; instead, one must be licensed to use their imagery. Thus, the ability to legally make derivative products from commercial satellite imagery is minimized.

Privacy concerns have been brought up by some who wish not to have their property shown from above. Google Maps responds to such concerns in their FAQ with the following statement: "*We understand your privacy concerns... The images that Google Maps displays are no different from what can be seen by anyone who flies over or drives by a specific geographic location.*"

Moving Images

In 2005 the Australian company Astrovision (ASX: HZG) announced plans to launch the first commercial geostationary satellite in the Asia-Pacific. It is intended to provide true color, real-time live satellite feeds, with down to 250 metres resolution over the entire Asia-Pacific region, from India to Hawaii and Japan to Australia. They were going to provide this content to users of 3G mobile phones, over Pay TV as a weather channel, and to corporate and government users.

Potential customers were excited by the possibilities offered, but they were unwilling (or, in government cases, generally unable) to sign contracts for a service that would not be delivered for 3–4 years (the length of time required to build and launch the satellite). AstroVision ran low on funds and was forced to shut down the program in 2006.

Aerial Photography

Aerial photography is the taking of photographs of the ground from an elevated/direct-down position. Usually the camera is not supported by a ground-based structure. Platforms for aerial photography include fixed-wing aircraft, helicopters, unmanned aerial vehicles (UAVs or "drones"), balloons, blimps and dirigibles, rockets, pigeons, kites, parachutes, stand-alone telescoping and vehicle-mounted poles. Mounted cameras may be triggered remotely or automatically; hand-held photographs may be taken by a photographer.

An aerial photograph using a drone of Westerheversand Lighthouse, Germany.

Air photo of a military target used to evaluate the effect of bombing.

Aerial photography should not be confused with air-to-air photography, where one or more aircraft are used as chase planes that "chase" and photograph other aircraft in flight.

History

Early History

Honoré Daumier, "Nadar élevant la Photographie à la hauteur de l'Art" (Nadar elevating Photography to Art), published in *Le Boulevard*, May 25, 1862.

Aerial photography was first practiced by the French photographer and balloonist Gaspard-Félix Tournachon, known as "Nadar", in 1858 over Paris, France. However, the photographs he produced no longer exist and therefore the earliest surviving aerial photograph is titled 'Boston, as the Eagle and the Wild Goose See It.' Taken by James Wallace Black and Samuel Archer King on October 13, 1860, it depicts Boston from a height of 630m.

Kite aerial photography was pioneered by British meteorologist E.D. Archibald in 1882. He used an explosive charge on a timer to take photographs from the air. Frenchman Arthur Batut began using kites for photography in 1888, and wrote a book on his methods in 1890. Samuel Franklin Cody developed his advanced 'Man-lifter War Kite' and succeeded in interesting the British War Office with its capabilities.

Antique postcard using kite photo technique.

The first use of a motion picture camera mounted to a heavier-than-air aircraft took place on April 24, 1909 over Rome in the 3:28 silent film short, *Wilbur Wright und seine Flugmaschine.*

World War I

Giza pyramid complex, photographed from Eduard Spelterini's balloon on November 21, 1904

The use of aerial photography rapidly matured during the war, as reconnaissance aircraft were equipped with cameras to record enemy movements and defences. At the start of the conflict, the usefulness of aerial photography was not fully appreciated, with reconnaissance being accomplished with map sketching from the air.

Germany adopted the first aerial camera, a Görz, in 1913. The French began the war with several squadrons of Blériot observation aircraft equipped with cameras for reconnaissance. The French Army developed procedures for getting prints into the hands of field commanders in record time.

Frederick Charles Victor Laws started aerial photography experiments in 1912 with

No.1 Squadron of the Royal Flying Corps (later No. 1 Squadron RAF), taking photographs from the British dirigible *Beta*. He discovered that vertical photos taken with 60% overlap could be used to create a stereoscopic effect when viewed in a stereoscope, thus creating a perception of depth that could aid in cartography and in intelligence derived from aerial images. The Royal Flying Corps recon pilots began to use cameras for recording their observations in 1914 and by the Battle of Neuve Chapelle in 1915, the entire system of German trenches was being photographed. In 1916 the Austro-Hungarian Monarchy made vertical camera axis aerial photos above Italy for map-making.

A German observation plane, the Rumpler Taube.

The first purpose-built and practical aerial camera was invented by Captain John Moore-Brabazon in 1915 with the help of the Thornton-Pickard company, greatly enhancing the efficiency of aerial photography. The camera was inserted into the floor of the aircraft and could be triggered by the pilot at intervals. Moore-Brabazon also pioneered the incorporation of stereoscopic techniques into aerial photography, allowing the height of objects on the landscape to be discerned by comparing photographs taken at different angles.

By the end of the war aerial cameras had dramatically increased in size and focal power and were used increasingly frequently as they proved their pivotal military worth; by 1918 both sides were photographing the entire front twice a day, and had taken over half a million photos since the beginning of the conflict. In January 1918, General Allenby used five Australian pilots from No. 1 Squadron AFC to photograph a 624 square miles (1,620 km^2) area in Palestine as an aid to correcting and improving maps of the Turkish front. This was a pioneering use of aerial photography as an aid for cartography. Lieutenants Leonard Taplin, Allan Runciman Brown, H. L. Fraser, Edward Patrick Kenny, and L. W. Rogers photographed a block of land stretching from the Turkish front lines 32 miles (51 km) deep into their rear areas. Beginning 5 January, they flew with a fighter escort to ward off enemy fighters. Using Royal Aircraft Factory BE.12 and Martinsyde airplanes, they not only overcame enemy air attacks, but also had to contend with 65 mph (105 km/h) winds, antiaircraft fire, and malfunctioning equipment to complete their task.

Commercial Aerial Photography

The first commercial aerial photography company in the UK was Aerofilms Ltd, founded by World War I veterans Francis Wills and Claude Graham White in 1919. The company soon expanded into a business with major contracts in Africa and Asia as well as in the UK. Operations began from the Stag Lane Aerodrome at Edgware, using the aircraft of the London Flying School. Subsequently the Aircraft Manufacturing Company (later the De Havilland Aircraft Company), hired an Airco DH.9 along with pilot entrepreneur Alan Cobham.

New York City 1930, aerial photograph of Fairchild Aerial Surveys Inc.

From 1921, Aerofilms carried out vertical photography for survey and mapping purposes. During the 1930s, the company pioneered the science of photogrammetry (mapping from aerial photographs), with the Ordnance Survey amongst the company's clients.

Another successful pioneer of the commercial use of aerial photography was the American Sherman Fairchild who started his own aircraft firm Fairchild Aircraft to develop and build specialized aircraft for high altitude aerial survey missions. One Fairchild aerial survey aircraft in 1935 carried unit that combined two synchronized cameras, and each camera having five six inch lenses with a ten-inch lenses and took photos from 23,000 feet. Each photo covered two hundred and twenty five square miles. One of its first government contracts was an aerial survey of New Mexico to study soil erosion. A year later, Fairchild introduced a better high altitude camera with nine-lens in one unit that could take a photo of 600 square miles with each exposure from 30,000 feet.

World War II

Sidney Cotton's Lockheed 12A, in which he made a high-speed reconnaissance flight in 1940.

In 1939 Sidney Cotton and Flying Officer Maurice Longbottom of the RAF were among the first to suggest that airborne reconnaissance may be a task better suited to fast, small aircraft which would use their speed and high service ceiling to avoid detection and interception. Although this seems obvious now, with modern reconnaissance tasks performed by fast, high flying aircraft, at the time it was radical thinking.

They proposed the use of Spitfires with their armament and radios removed and replaced with extra fuel and cameras. This led to the development of the Spitfire PR variants. Spitfires proved to be extremely successful in their reconnaissance role and there were many variants built specifically for that purpose. They served initially with what later became No. 1 Photographic Reconnaissance Unit (PRU). In 1928, the RAF developed an electric heating system for the aerial camera. This allowed reconnaissance aircraft to take pictures from very high altitudes without the camera parts freezing. Based at RAF Medmenham, the collection and interpretation of such photographs became a considerable enterprise.

Cotton's aerial photographs were far ahead of their time. Together with other members of the 1 PRU, he pioneered the techniques of high-altitude, high-speed stereoscopic photography that were instrumental in revealing the locations of many crucial military and intelligence targets. According to R.V. Jones, photographs were used to establish the size and the characteristic launching mechanisms for both the V-1 flying bomb and the V-2 rocket. Cotton also worked on ideas such as a prototype specialist reconnaissance aircraft and further refinements of photographic equipment. At the peak, the British flew over 100 reconnaissance flights a day, yielding 50,000 images per day to interpret. Similar efforts were taken by other countries.

Uses

Abalone point .. Irvine Cove, Laguna Beach an example of low-altitude aerial photography

Aerial photography is used in cartography (particularly in photogrammetric surveys, which are often the basis for topographic maps), land-use planning, archaeology, movie production, environmental studies, power line inspection, surveillance, commercial advertising, conveyancing, and artistic projects. An example of how aerial photography is used in the field of archaeology is the mapping project done at the site Angkor Borei in Cambodia from 1995-1996. Using aerial photography, archaeologists were able to identify archaeological features, including 112 water features (reservoirs, artificially constructed pools and natural ponds) within the walled site of Angkor Borei. In the United States, aerial photographs are used in many Phase I Environmental Site Assessments for property analysis.

Platforms

Aircraft

Full-size, manned aircraft are prohibited from flights under 1000 fcct, over congested areas and 500 feet above more sparsely populated locations.

Radio-controlled Model Aircraft

A drone carrying a camera for aerial photography

Two drones that can be used to take aerial photographs

Advances in radio controlled models have made it possible for model aircraft to conduct low-altitude aerial photography. This had benefited real-estate advertising, where commercial and residential properties are the photographic subject when in 2014 the US Federal Communications Commission, issued an order banning the use of "Drones" in any commercial application related to photographs for use in real estate advertisement's. Small scale model aircraft offer increased photographic access to these previously restricted areas. Miniature vehicles do not replace full size aircraft, as full size aircraft are capable of longer flight times, higher altitudes, and greater equipment payloads. They are, however, useful in any situation in which a full-scale aircraft would be dangerous to operate. Examples would include the inspection of transformers atop power transmission lines and slow, low-level flight over agricultural fields, both of which can be accomplished by a large-scale radio controlled helicopter. Professional-grade, gyroscopically stabilized camera platforms are available for use under such a model; a large model helicopter with a 26cc gasoline engine can hoist a payload of approximately seven kilograms (15 lbs). In addition to gyroscopically stabilized footage, the use of RC copters as reliable aerial photography tools increased with the integration of FPV (first-person-view) technology. Many radio-controlled aircraft are now capable of utilizing Wi-Fi to stream live video from the aircraft's camera back to the pilot's ground station.

Regulations

Australia

In Australia Civil Aviation Safety Regulation 101 (CASR 101) allows for commercial use of radio control aircraft. Under these regulations radio controlled unmanned aircraft for commercial are referred to as Unmanned Aircraft Systems (UAS), where as radio controlled aircraft for recreational purposes are referred to as model aircraft. Under CASR 101, businesses/persons operating radio controlled aircraft commercially are required to hold an operator certificate, just like manned aircraft operators. Pilots of radio controlled aircraft operating commercially are also required to be licensed by the Civil Aviation Safety Authority (CASA). Whilst a small UAS and model aircraft may actually be identical, unlike model aircraft, a UAS may enter controlled airspace with approval, and operate within close proximity to an aerodrome.

Due to a number of illegal operators in Australia making false claims of being approved, CASA maintains and publishes a list of approved UAS operators.

However CASA has modified the regulations and from the 29th of September 2016 drones under 2kg may be operated for commercial purposes. This makes a lot of sense in creating a more competitive market place for basic services such as aerial photography in the real estate industry and farming.[*according to whom?*]

United States

Recent (2006) FAA regulations grounding all commercial RC model flights have been upgraded to require formal FAA certification before permission is granted to fly at any altitude in the US.

June 25, 2014, The FAA, in ruling 14 CFR Part 91 [Docket No. FAA–2014–0396] "Interpretation of the Special Rule for Model Aircraft", banned the commercial use of unmanned aircraft over U.S. airspace. On September 26, 2014, the FAA began granting the right to use drones in aerial filmmaking. Operators are required to be licensed pilots and must keep the drone in view at all times. Drones cannot be used to film in areas where people might be put at risk.

On February 14, 2012, the President signed into law the FAA Modernization and Reform Act of 2012 (Pub. L. 112–95) (the Act), which established, in Section 336, a special rule for model aircraft. In Section 336, Congress confirmed the FAA's long-standing position that model aircraft are aircraft. Under the terms of the Act, a model aircraft is defined as "an unmanned aircraft" that is "(1) capable of sustained flight in the atmosphere; (2) flown within visual line of sight of the person operating the aircraft; and (3) flown for hobby or recreational purposes."

Because anything capable of being viewed from a public space is considered outside the realm of privacy in the United States, aerial photography may legally document features and occurrences on private property.

The FAA can pursue enforcement action against persons operating model aircraft who endanger the safety of the national airspace system. Public Law 112–95, section 336(b).

Types

Oblique

Oblique Aerial Photo

Photographs taken at an angle are called *oblique photographs*. If they are taken from a low angle earth surface–aircraft, they are called *low oblique* and photographs taken from a high angle are called *high* or *steep oblique*.

An aerial photographer prepares continuous oblique shooting in a Cessna 206

Vertical

Vertical Orientation Aerial Photo

Vertical photographs are taken straight down. They are mainly used in photogrammetry and image interpretation. Pictures that will be used in photogrammetry are traditionally taken with special large format cameras with calibrated and documented geometric properties.

Combinations

Aerial photographs are often combined. Depending on their purpose it can be done in several ways, of which a few are listed below.

- Panoramas can be made by stitching several photographs taken with one hand held camera.
- In pictometry five rigidly mounted cameras provide one vertical and four low oblique pictures that can be used together.
- In some digital cameras for aerial photogrammetry images from several imaging elements, sometimes with separate lenses, are geometrically corrected and combined to one image in the camera.

Orthophotos

Vertical photographs are often used to create orthophotos, alternatively known as orthophotomaps, photographs which have been geometrically "corrected" so as to be usable as a map. In other words, an orthophoto is a simulation of a photograph taken from an infinite distance, looking straight down to nadir. Perspective must obviously be removed, but variations in terrain should also be corrected for. Multiple geometric transformations are applied to the image, depending on the perspective and terrain corrections required on a particular part of the image.

Orthophotos are commonly used in geographic information systems, such as are used by mapping agencies (e.g. Ordnance Survey) to create maps. Once the images have been aligned, or "registered", with known real-world coordinates, they can be widely deployed.

Large sets of orthophotos, typically derived from multiple sources and divided into "tiles" (each typically 256 x 256 pixels in size), are widely used in online map systems such as Google Maps. OpenStreetMap offers the use of similar orthophotos for deriving new map data. Google Earth overlays orthophotos or satellite imagery onto a digital elevation model to simulate 3D landscapes.

Aerial Video

With advancements in video technology, aerial video is becoming more popular. Orthogonal video is shot from aircraft mapping pipelines, crop fields, and other points of interest. Using GPS, video may be embedded with meta data and later synced with a video mapping program.

This "Spatial Multimedia" is the timely union of digital media including still photography, motion video, stereo, panoramic imagery sets, immersive media constructs, audio, and other data with location and date-time information from the GPS and other location designs.

Aerial videos are emerging Spatial Multimedia which can be used for scene understanding and object tracking. The input video is captured by low flying aerial platforms and typically consists of strong parallax from non-ground-plane structures. The integration of digital video, global positioning systems (GPS) and automated image processing will improve the accuracy and cost-effectiveness of data collection and reduction. Several different aerial platforms are under investigation for the data collection.

Prismatic Compass (Surveying)

A prismatic compas is a navigation and surveying instrument which is extensively used for determining course, waypoints (an endpoint of the leg of a course) and direction, and for calculating bearings of survey lines and included angles between them. Compass surveying is a type of surveying in which the directions of surveying lines are determined with a magnetic compass, and the length of the surveying lines are measured with a tape or chain or laser range finder. The compass is generally used to run a traverse line. The compass calculates bearings of lines with respect to magnetic north. The included angles can then be calculated using suitable formulas in case of clockwise and anti-clockwise traverse respectively. For each survey line in the traverse, surveyors take two bearings that is fore bearing and back bearing which should exactly differ by 180° if local attraction is negligible. The name *Prismatic compass* is given to it because it essentially consists of a prism which is used for taking observations more accurately.

Least Count

Least count means the minimum value that an instrument can read which is 15 minutes in case of prismatic compass. It means compass can read only those observations which are multiples of 30 minutes, $5^\circ 25'$, $15^\circ 55'$, $35^\circ 45'$ 30<sup

Bearings

The compass calculates the bearings in whole circle bearing system which determines the angle which the survey line makes with the magnetic north in the clockwise direction. The included angles can be calculated by the formulas F-P ±180 in case of anti-clocwise traverse and P-F ±180 in case of clockwise traverse, where 'F' is the fore bearing of forward line in the direction of survey work and 'P' is the fore bearing of previous line.

Parts

The essential parts of the prismatic compass are listed below:-

- Magnetic needle
- Eye vane
- Eye slit
- Eye hole
- Mirror (adjustable)
- Glass cover
- Graduated ring
- Lifting lever
- Lifting pin
- Metal box (85–100 mm dia)
- Focusing stud
- Object vane
- Sun glasses
- Sliding arrangement for mirror
- Horse hair

Geomorphometry

Geomorphometry is the science of quantitative land surface analysis. It gathers various mathematical, statistical and image processing techniques that can be used to quantify morphological, hydrological, ecological and other aspects of a land surface. Common synonyms for geomorphometry are geomorphological analysis, terrain morphometry or terrain analysis and land surface analysis.

In simple terms, geomorphometry aims at extracting (land) surface parameters (morphometric, hydrological, climatic etc.) and objects (watersheds, stream networks, landforms etc.) using input digital land surface model (also known as digital elevation model, DEM) and parameterization software. Extracted surface parameters and objects can then be used, for example, to improve mapping and modelling of soils, vegetation, land use, geomorphological and geological features and similar.

Although geomorphometry started with ideas of Brisson (1808) and Gauss (1827), the field did not evolve much until the construction of the first DEM. With the rapid increase of sources of DEMs today (and especially due to the Shuttle Radar Topography Mission and LIDAR-based projects), extraction of land surface parameters is becoming more and more attractive to numerous fields ranging from precision agriculture, soil-landscape modelling, climatic and hydrological applications to urban planning, education and space research. The topography of almost all Earth has been today sampled or scanned, so that DEMs are available at resolutions of 100 m or better at global scale. Land surface parameters are today successfully used for both stochastic and process-based modelling, the only remaining issue being the level of detail and vertical accuracy of the DEM.

Photogrammetry

Photogrammetry is the science of making measurements from photographs, especially for recovering the exact positions of surface points. It may also be used to recover the motion pathways of designated reference points on any moving object, on its components, and in the immediately adjacent environment. Photogrammetric analysis may be applied to one photograph, or may use high-speed photography and remote sensing to detect, measure and record complex 2-D and 3-D motion fields. Photogrammetry feeds measurements from remote sensing and the results of imagery analysis into computational models in an attempt to successively estimate, with increasing accuracy, the actual, 3-D relative motions within the researched field.

Low altitude aerial photograph for use in photogrammetry. Location: Three Arch Bay, Laguna Beach, CA.

Its applications include satellite tracking of the relative positioning alterations in all Earth environments (e.g. tectonic motions etc.), research on the swimming of fish, of bird or insect flight, and other relative motion processes (International Society for Photogrammetry and Remote Sensing). The quantitative results of photogrammetry are then used to guide and match the results of computational models of the natural systems. They help to invalidate or confirm new theories, to design novel vehicles or new methods for predicting or/and controlling the consequences of earthquakes, tsunamis or other events, or to understand the flow of fluids next to solid structures and many other processes.

Photogrammetry is as old as modern photography, dating to the mid-19th century. Its detection component has been emerging from radiolocation, multilateration and radiometry. Its 3-D positioning estimative component (based on modeling) employs methods related to triangulation, trilateration and multidimensional scaling.

In the simplest example, the distance between two points that lie on a plane parallel to the photographic image plane, can be determined by measuring their distance on the image, if the scale (s) of the image is known. This is done by multiplying the measured distance by $1/s$.

Photogrammetric Methods

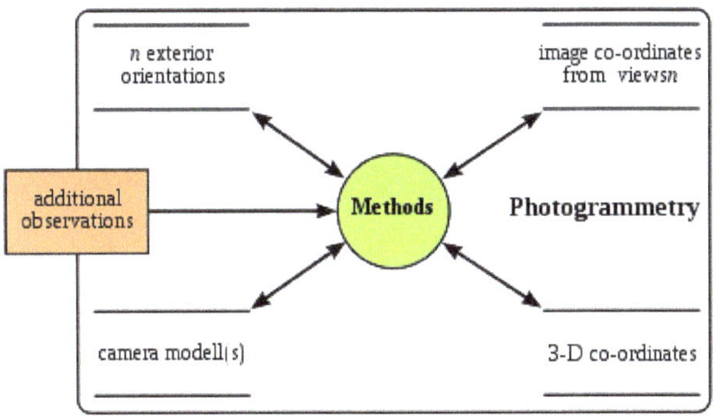

Georg Wiora's data model of photogrammetry

Photogrammetry has been defined by the American Society for Photogrammetry and Remote Sensing (ASPRS) as the art, science, and technology of obtaining reliable information about physical objects and the environment through processes of recording, measuring and interpreting photographic images and patterns of recorded radiant electromagnetic energy and other phenomena.

Photogrammetry uses methods from many disciplines, including optics and projective geometry. Digital image capturing and photogrammetric processing includes several well defined stages, which allow to generate 2D or 3D digital models of the object as an end product. The data model on the right shows what type of information can go into and come out of photogrammetric methods.

The *3-D co-ordinates* define the locations of object points in the 3-D space. The *image co-ordinates* define the locations of the object points' images on the film or an electronic imaging device. The *exterior orientation* of a camera defines its location in space and its view direction. The *inner orientation* defines the geometric parameters of the imaging process. This is primarily the focal length of the lens, but can also include the description of lens distortions. Further *additional observations* play an important role: With *scale bars*, basically a known distance of two points in space, or known *fix points*, the connection to the basic measuring units is created.

Each of the four main variables can be an *input* or an *output* of a photogrammetric method.

Algorithms for photogrammetry typically attempt to minimize the sum of the squares of errors over the coordinates and relative displacements of the reference points. This minimization is known as bundle adjustment and is often performed using the Levenberg–Marquardt algorithm.

Stereophotogrammetry

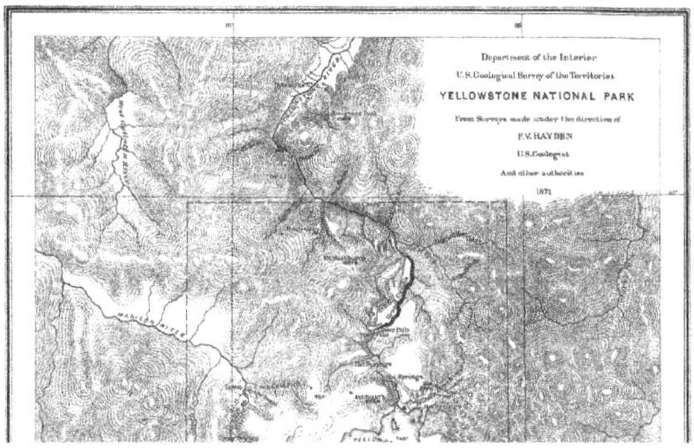

The stereophotogrammetry technology Rapid 3D Mapping applied on the Royal Castle of Sweden.

A special case, called stereophotogrammetry, involves estimating the three-dimensional coordinates of points on an object employing measurements made in two or more photographic images taken from different positions. Common points are identified on each image. A line of sight (or ray) can be constructed from the cam-era location to the point on the object. It is the intersection of these rays (triangulation) that determines the three-dimensional location of the point. More sophisticated algorithms can exploit other information about the scene that is known *a priori*, for example symmetries, in some cases allowing reconstructions of 3-D coordinates from only one camera position. Stereophotogrammetry is emerging as a robust non-contacting measurement technique to determine dynamic characteristics and mode shapes of non-rotating and rotating structures.

Integration

Photogrammetric data with a dense range data in which scanners complement each other. Photogrammetry is more accurate in the x and y direction while range data are generally more accurate in the z direction. This range data can be supplied by techniques like LiDAR, laser scanners (using time of flight, triangulation or interferometry), white-light digitizers and any other technique that scans an area and returns x, y, z coordinates for multiple discrete points (commonly called "point clouds"). Photos can clearly define the edges of buildings when the point cloud footprint can not. It is beneficial to incorporate the advantages of both systems and integrate them to create a better product.

A 3-D visualization can be created by georeferencing the aerial photos and LiDAR data in the same reference frame, orthorectifying the aerial photos, and then draping the orthorectified images on top of the LiDAR grid. It is also possible to create digital terrain models and thus 3-D visualisations using pairs (or multiples) of aerial photographs or satellite (e.g. SPOT satellite imagery). Techniques such as adaptive least squares stereo matching are then used to produce a dense array of correspondences which are transformed through a camera model to produce a dense array of x, y, z data which can be used to produce digital terrain model and orthoimage products. Systems which use these techniques, e.g. the ITG system, were developed in the 1980s and 1990s but have since been supplanted by LiDAR and radar-based approaches, although these techniques may still be useful in deriving elevation models from old aerial photographs or satellite images.

Applications

Video of a 3-D model of Horatio Nelson bust in Monmouth Museum, produced using photogrammetry.

Photogrammetry is used in different fields, such as topographic mapping, architecture, engineering, manufacturing, quality control, police investigation, and geology, as well as by archaeologists to quickly produce plans of large or complex sites and by meteorologists as a way to determine the actual wind speed of a tornado where objective weather data cannot be obtained.

It is also used to combine live action with computer-generated imagery in movies post-production; *The Matrix* is a good example of the use of photogrammetry in film (details are given in the DVD extras). Photogrammetry was used extensively to create photorealistic environmental assets for video games including *The Vanishing of Ethan Carter* as well as EA DICE's *Star Wars Battlefront*.

A somewhat similar application is the scanning of objects to automatically make 3D models of them. Some programs like RealityCapture, Acute3D's Smart3DCapture, now part of Bentley Systems and renamed ContextCapture, Pix4Dmapper, Photoscan, 123D Catch, Bundler toolkit, PIXDIM, and Photosketch have been made to allow people to quickly make 3D models using this photogrammetry method. It should be noted though that the produced model often still contains gaps, so additional cleanup with software like MeshLab, netfabb or MeshMixer is often still necessary.

Photogrammetry is also commonly employed in collision engineering, especially with automobiles. When litigation for accidents occurs and engineers need to determine the exact deformation present in the vehicle, it is common for several years to have passed and the only evidence that remains is accident scene photographs taken by the police. Photogrammetry is used to determine how much the car in question was deformed, which relates to the amount of energy required to produce that deformation. The energy can then be used to determine important information about the crash (such as the velocity at time of impact).

Software

There exist many software packages for photogrammetry; comparison of photogrammetry software.

Digital Elevation Model

A digital elevation model (DEM) is a digital model or 3D representation of a terrain's surface — commonly for a planet (including Earth), moon, or asteroid — created from terrain elevation data.

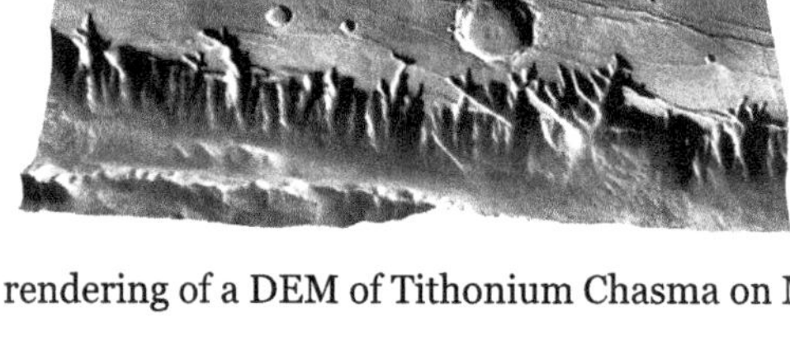

3D rendering of a DEM of Tithonium Chasma on Mars

Terminology

There is no universal usage of the terms *digital elevation model* (DEM), *digital terrain model* (DTM) and *digital surface model* (DSM) in scientific literature. In most cases the term *digital surface model* represents the earth's surface and includes all objects on it. In contrast to a DSM, the *digital terrain model* (DTM) represents the bare ground surface without any objects like plants and buildings.

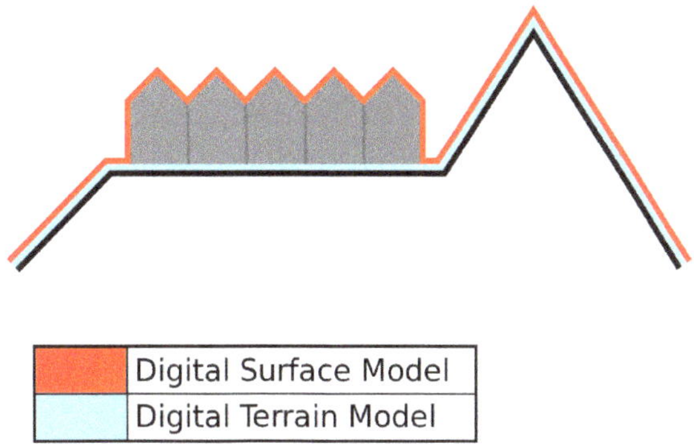

Surfaces represented by a Digital Surface Model include buildings and other objects. Digital Terrain Models represent the bare ground.

DEM is often used as a generic term for DSMs and DTMs, only representing height information without any further definition about the surface. Other definitions equalise the terms DEM and DTM, or define the DEM as a subset of the DTM, which also represents other morphological elements. There are also definitions which equalise the terms DEM and DSM. On the Web definitions can be found which define DEM as a regularly spaced GRID and a DTM as a three-dimensional model (TIN). Most of the data providers (USGS, ERSDAC, CGIAR, Spot Image) use the term DEM as a generic term for DSMs and DTMs. All datasets which are captured with satellites, airplanes or other flying platforms are originally DSMs (like SRTM or the ASTER GDEM). It is possible to compute a DTM from high resolution DSM datasets with complex algorithms (Li *et al.*, 2005). In the following the term DEM is used as a generic term for DSMs and DTMs.

Types of DEM

Heightmap of Earth's surface (including water and ice) in equirectangular projection, normalized as 8-bit grayscale, where lighter values indicate higher elevation

A DEM can be represented as a raster (a grid of squares, also known as a heightmap when representing elevation) or as a vector-based triangular irregular network (TIN). The TIN DEM dataset is also referred to as a primary (measured) DEM, whereas the Raster DEM is referred to as a secondary (computed) DEM. The DEM could be acquired through techniques such as photogrammetry, lidar, IfSAR, land surveying, etc. (Li et al. 2005). DEMs are commonly built using data collected using remote sensing techniques, but they may also be built from land surveying. DEMs are used often in geographic information systems, and are the most common basis for digitally produced relief maps. While a DSM may be useful for landscape modeling, city modeling and visualization applications, a DTM is often required for flood or drainage modeling, land-use studies, geological applications, and other applications.

Production

Relief map Sierra Nevada

Mappers may prepare digital elevation models in a number of ways, but they frequently use remote sensing rather than direct survey data. One powerful technique for generating digital elevation models is interferometric synthetic aperture radar where two passes of a radar satellite (such as RADARSAT-1 or TerraSAR-X or Cosmo SkyMed), or a single pass if the satellite is equipped with two antennas (like the SRTM instrumentation), collect sufficient data to generate a digital elevation map tens of kilometers on a side with a resolution of around ten meters. Other kinds of stereoscopic pairs can be employed using the digital image correlation method, where two optical images are acquired with different angles taken from the same pass of an airplane or an Earth Observation Satellite (such as the HRS instrument of SPOT5 or the VNIR band of ASTER).

The SPOT 1 satellite (1986) provided the first usable elevation data for a sizeable portion of the planet's landmass, using two-pass stereoscopic correlation. Later, further data were provided by the European Remote-Sensing Satellite (ERS, 1991) using the same method, the Shuttle Radar Topography Mission (SRTM, 2000) using single-pass SAR and the Advanced Spaceborne Thermal Emission and Reflection Radiometer (ASTER, 2000) instrumentation on the Terra satellite using double-pass stereo pairs.

The HRS instrument on SPOT 5 has acquired over 100 million square kilometers of stereo pairs.

Older methods of generating DEMs often involve interpolating digital contour maps that may have been produced by direct survey of the land surface. This method is still used in mountain areas, where interferometry is not always satisfactory. Note that contour line data or any other sampled elevation datasets (by GPS or ground survey) are not DEMs, but may be considered digital terrain models. A DEM implies that elevation is available continuously at each location in the study area.

The quality of a DEM is a measure of how accurate elevation is at each pixel (absolute accuracy) and how accurately is the morphology presented (relative accuracy). Several factors play an important role for quality of DEM-derived products:

- terrain roughness;
- sampling density (elevation data collection method);
- grid resolution or pixel size;
- interpolation algorithm;
- vertical resolution;
- terrain analysis algorithm;
- Reference 3D products include quality masks that give information on the coastline, lake, snow, clouds, correlation etc.

Methods for Obtaining Elevation Data Used to Create DEMs

Gatewing X100 unmanned aerial vehicle

- Lidar
- Stereo photogrammetry from aerial surveys
- Structure from motion / Multi-view stereo applied to aerial photography
- Block adjustment from optical satellite imagery
- Interferometry from radar data
- Real Time Kinematic GPS
- Topographic maps

- Theodolite or total station
- Doppler radar
- Focus variation
- Inertial surveys
- Surveying and mapping drones
- Range imaging

Uses

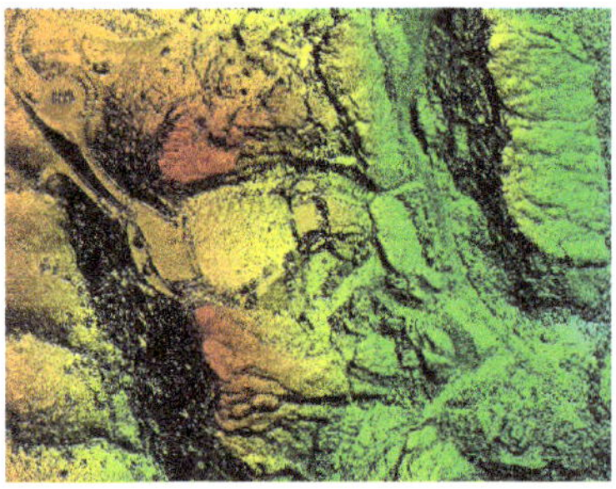

Digital Elevation Model - Red Rocks Amphitheater, Colorado obtained using an UAV (DroneMapper)

Bezmiechowa airfield 3D Digital Surface Model obtained using Pteryx UAV flying 200 m above hilltop

Digital Surface Model of motorway interchange construction site. Note that tunnels are closed.

Common uses of DEMs include:

- Extracting terrain parameters for geomorphology

- Modeling water flow for hydrology or mass movement (for example avalanches and landslides)
- Modeling soils wetness with Cartographic Depth to Water Indexes (DTW-index)
- Creation of relief maps
- Rendering of 3D visualizations.
- 3D flight planning and TERCOM
- Creation of physical models (including raised relief maps)
- Rectification of aerial photography or satellite imagery
- Reduction (terrain correction) of gravity measurements (gravimetry, physical geodesy)
- Terrain analysis in geomorphology and physical geography
- Geographic Information Systems (GIS)
- Engineering and infrastructure design
- Satellite navigation (for example GPS and GLONASS)
- Line-of-sight analysis
- Base mapping
- Flight simulation
- Precision farming and forestry
- Surface analysis
- Intelligent transportation systems (ITS)
- Auto safety / Advanced Driver Assistance Systems (ADAS)
- Archaeology

Sources

A free DEM of the whole world called GTOPO30 (30 arcsecond resolution, approx. 1 km) is available, but its quality is variable and in some areas it is very poor. A much higher quality DEM from the Advanced Spaceborne Thermal Emission and Reflection Radiometer (ASTER) instrument of the Terra satellite is also freely available for 99% of the globe, and represents elevation at 30 meter resolution. A similarly high resolution was previously only available for the United States territory under the Shuttle Radar Topography Mission (SRTM) data, while most of the rest of the planet was only covered in a 3 arc-second resolution (around 90 meters). The limitation with the GTOPO30 and SRTM datasets is that they cover continental landmasses only, and SRTM does not cover the polar regions and has mountain and desert no data (void) areas. SRTM data, being derived from radar, represents the elevation of the first-reflected surface — quite often tree tops. So, the data are not necessarily representative of the ground surface, but the top of whatever is first encountered by the radar.

Submarine elevation (known as bathymetry) data is generated using ship-mounted depth soundings. When land topography and bathymetry is combined, a truly Global Relief Model is obtained. The SRTM30Plus dataset (used in NASA World Wind) attempts to combine GTOPO30, SRTM and bathymetric data to produce a truly global elevation model. The Earth2014 global topography and relief model provides layered topography grids at 1 arc-minute resolution. Other than SRTM30plus, Earth2014 provides information on ice-sheet heights and bedrock (that is, topography below the ice) over Antarctica and Greenland. Another global model is Global Multi-resolution Terrain Elevation Data 2010 (GMTED2010) with 7.5 arc second resolution. It is based on SRTM data and combines other data outside SRTM coverage. A novel global DEM of postings lower than 12 m and a height accuracy of less than 2 m is expected from the TanDEM-X satellite mission which started in July 2010.

The most common grid (raster) spacing is between 50 and 500 meters. In gravimetry e.g., the primary grid may be 50 m, but is switched to 100 or 500 meters in distances of about 5 or 10 kilometers.

Since 2002, the HRS instrument on SPOT 5 has acquired over 100 million square kilometers of stereo pairs used to produce a DTED2 format DEM (with a 30-meter posting) DEM format DTED2 over 50 million km^2. The radar satellite RADARSAT-2 has been used by MacDonald, Dettwiler and Associates Ltd. to provide DEMs for commercial and military customers.

In 2014, acquisitions from radar satellites TerraSAR-X and TanDEM-X will be available in the form of a uniform global coverage with a resolution of 12 meters.

Many national mapping agencies produce their own DEMs, often of a higher resolution and quality, but frequently these have to be purchased, and the cost is usually prohibitive to all except public authorities and large corporations. DEMs are often a product of national lidar dataset programs.

Free DEMs are also available for Mars: the MEGDR, or Mission Experiment Gridded Data Record, from the Mars Global Surveyor's Mars Orbiter Laser Altimeter (MOLA) instrument; and NASA's Mars Digital Terrain Model (DTM).

OpenTopography is a community access resource of high-resolution, Earth science-oriented, topography data, and related tools and resources that warehouses a great deal of DEM data. OpenTopography is based at the San Diego Supercomputer Center at the

University of California, San Diego and is operated in collaboration with colleagues in the School of Earth and Space Exploration at Arizona State University. Core operational support for OpenTopography comes from the National Science Foundation Earth Sciences: Instrumentation and Facilities Program (EAR/IF) and the Office of Cyberinfrastructure.

United States

The US Geological Survey produces the National Elevation Dataset, a seamless DEM for the United States, Hawaii and Puerto Rico based on 7.5' topographic mapping. As of the beginning of 2006, this replaces the earlier DEM tiled format (one DEM per USGS topographic map).

OpenTopography is a U.S.-based community access source of a large amount of high-resolution topography data for the U.S.

Bathymetry

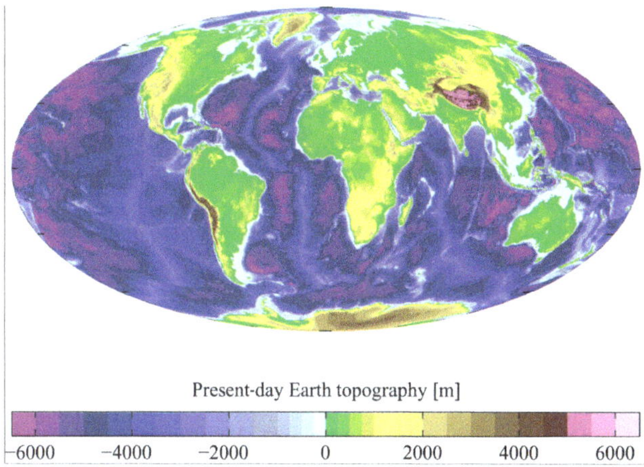

Present day Earth bathymetry (and altimetry). Data from the National Geophysical Data Center's TerrainBase Digital Terrain Model.

Bathymetry is the study of underwater depth of lake or ocean floors. In other words, bathymetry is the underwater equivalent to hypsometry or topography. The name comes from Greek βαθὺς (bathus), "deep", and μέτρον (metron), "measure". Bathymetric (or hydrographic) charts are typically produced to support safety of surface or sub-surface navigation, and usually show seafloor relief or terrain as contour lines (called depth contours or isobaths) and selected depths (soundings), and typically also provide surface navigational information. Bathymetric maps (a more general term where navigational safety is not a concern) may also use a Digital Terrain Model and artificial illumination techniques to illustrate the depths being portrayed. The global bathymetry is sometimes combined with topography data to yield a Global Relief Model. Paleobathymetry is the study of past underwater depths.

Measurement

First printed map of oceanic bathymetry, produced with data from USS *Dolphin* (1853).

Originally, bathymetry involved the measurement of ocean depth through depth sounding. Early techniques used pre-measured heavy rope or cable lowered over a ship's side. This technique measures the depth only a singular point at a time, and is therefore inefficient. It is also subject to movements of the ship and currents moving the line out of true and therefore is inaccurate.

The data used to make bathymetric maps today typically comes from an echosounder (sonar) mounted beneath or over the side of a boat, "pinging" a beam of sound downward at the seafloor or from remote sensing LIDAR or LADAR systems. The amount of time it takes for the sound or light to travel through the water, bounce off the seafloor, and return to the sounder informs the equipment of the distance to the seafloor. LIDAR/LADAR surveys are usually conducted by airborne systems.

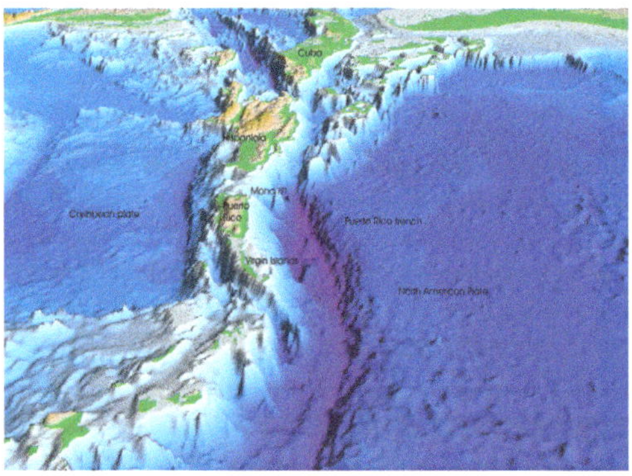

The seafloor topography near the Puerto Rico Trench

Starting in the early 1930s, single-beam sounders were used to make bathymetry maps. Today, multibeam echosounders (MBES) are typically used, which use hundreds of very narrow adjacent beams arranged in a fan-like swath of typically 90 to 170 degrees across. The tightly packed array of narrow individual beams provides very high angular resolution and accuracy. In general a wide swath, which is depth dependent, allows a boat to map more seafloor in less time than a single-beam echosounder by making fewer passes. The beams update many times per second (typically 0.1-50 Hz depending on water depth), allowing faster boat speed while maintaining 100% coverage of the seafloor. Attitude sensors allow for the correction of the boat's roll and pitch on the ocean surface, and a gyrocompass provides accurate heading information to correct for vessel yaw. (Most modern MBES systems use an integrated motion-sensor and position system that measures yaw as well as the other dynamics and position.) A boat-mounted Global Positioning System (GPS) (or other Global Navigation Satellite System (GNSS)) positions the soundings with respect to the surface of the earth. Sound speed profiles (speed of sound in water as a function of depth) of the water column correct for refraction or "ray-bending" of the sound waves owing to non-uniform water column characteristics such as temperature, conductivity, and pressure. A computer system processes all the data, correcting for all of the above factors as well as for the angle of each individual beam. The resulting sounding measurements are then processed either manually, semi-automatically or automatically (in limited circumstances) to produce a map of the area. As of 2010 a number of different outputs are generated, including a sub-set of the original measurements that satisfy some conditions (e.g., most representative likely soundings, shallowest in a region, etc.) or integrated Digital Terrain Models (DTM) (e.g., a regular or irregular grid of points connected into a surface). Historically, selection of measurements was more common in hydrographic applications while DTM construction was used for engineering surveys, geology, flow modeling, etc. Since ca. 2003-2005, DTMs have become more accepted in hydrographic practice.

Satellites are also used to measure bathymetry. Satellite radar maps deep-sea topography by detecting the subtle variations in sea level caused by the gravitational pull of undersea mountains, ridges, and other masses. On average, sea level is higher over mountains and ridges than over abyssal plains and trenches.

In the United States the United States Army Corps of Engineers performs or commissions most surveys of navigable inland waterways, while the National Oceanic and Atmospheric Administration (NOAA) performs the same role for ocean waterways. Coastal bathymetry data is available from NOAA's National Geophysical Data Center (NGDC), which is now merged into National Centers for Environmental Information. Bathymetric data is usually referenced to tidal vertical datums. For deep-water bathymetry, this is typically Mean Sea Level (MSL), but most data used for nautical charting is referenced to Mean Lower Low Water (MLLW) in American surveys, and Lowest Astronomical Tide (LAT) in other countries. Many other datums are used in practice, depending on the locality and tidal regime.

Occupations or careers related to bathymetry include the study of oceans and rocks and minerals on the ocean floor, and the study of underwater earthquakes or volcanoes. The taking and analysis of bathymetric measurements is one of the core areas of modern hydrography, and a fundamental component in ensuring the safe transport of goods worldwide.

Global Relief Model

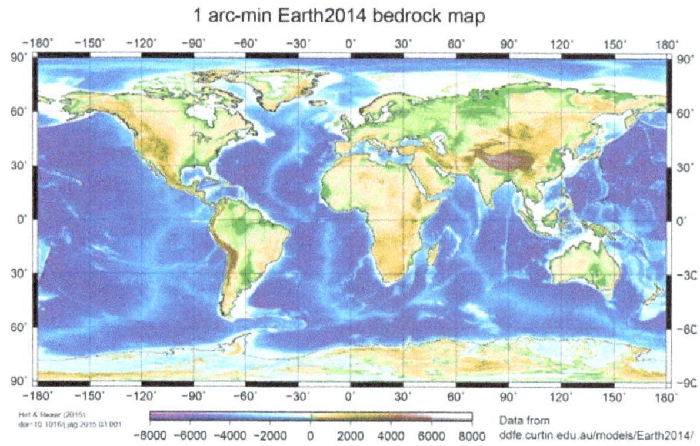

Example of a global relief model: Earth2014 bedrock layer (topography over land, bathymetry over oceans and major lakes, sub-ice-topography over ice-shields)

Terminology

A global relief model, sometimes also denoted as global topography model or composite model, combines digital elevation model (DEM) data over land with digital bathymetry model (DBM) data over water-covered areas (oceans, lakes) to describe Earth's relief. A relief model thus shows how Earth's surface would look like in the absence of water or ice masses.

The relief is represented by a set of heights (elevations or depths) that refer to some height reference surface, often the mean sea level or the geoid. Global relief models are used for a variety of applications including visualisation, geological, geomorphological and geophysical analyses, gravity field modelling as well as geo-statistics.

Measurement

Global relief models are always based on combinations of data sets from different remote sensing techniques. This is because there is no single remote sensing technique that would allow measurement of the relief both over dry and water-covered areas. Elevation data over land is often obtained from LIDAR or inSAR measurements, while bathymetry is acquired based on SONAR and altimetry. Global relief models may also contain elevations of the bedrock (sub-ice-topography) below the ice shields of Greenland and Antarctica. Ice sheet thickness, mostly measured through ice-penetrating RADAR, is subtracted from the ice surface heights to reveal the bedrock.

Spatial Resolution

While digital elevation models describe Earth's land topography often with 1 to 3 arc-second resolution (e.g., from the SRTM or ASTER missions), the global bathymetry (e.g., SRTM30_PLUS) is known to a much lesser spatial resolution in the kilometre-range. The same holds true for models of the bedrock of Antarctica and Greenland. Therefore, global relief models are often constructed at 1 arc-min resolution (corresponding to about 1.8 km postings), Some products such as the 30 and 15 arc-sec resolution SRTM30_PLUS/ SRTM15_PLUS grids offer higher resolution to adequately represent SONAR depth measurements where available.

Public Data Sets

Data sets produced and released to the public include Earth2014, SRTM30_PLUS and ETOPO1.

Earth2014 (2015)

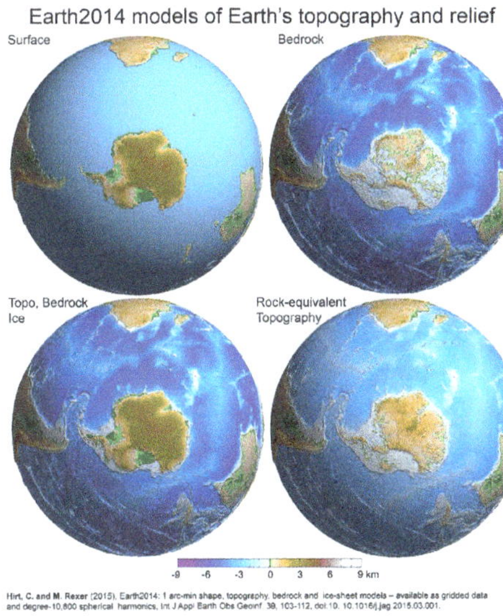

Four different topography layers of the Earth2014 model. Clockwise from top left: (1) Earth's surface, (2) bedrock, (3) rock-equivalent topography, (4) bathymetry and ice surface

The most recent global relief model is Earth2014, developed at Curtin University (Western Australia) and TU Munich (Germany). Earth2014 provides sets of 1 arc-min resolution global grids of Earth's relief in different representations based on the 2013 releases of bedrock and ice-sheet data over Antarctica (Bedmap2) and Greenland (Greenland Bedrock Topography), the 2013 SRTM_30PLUS bathymetry and 2008 SRTM V4.1 SRTM land topography.

Earth2014 provides five different layers of height data, comprising Earth's surface (lower interface of the atmosphere), topography and bathymetry of the oceans and major lakes, topography, bathymetry and bedrock, ice-sheet thicknesses and rock-equivalent topography. The Earth2014 global grids are provided as heights relative to the mean sea level, and as planetary radii relative to the centre of Earth, which show the shape of the Earth.

SRTM30_PLUS (2014)

SRTM30_PLUS is a combined bathymetry and topography model produced by Scripps Institution of Oceanography (California). The version 15_PLUS comes at 0.25 arc-min resolution (about 450 m postings), while the 30_PLUS version offers 0.5 arc-min (900 m) resolution. The bathymetric data in SRTM30_PLUS stems from depth soundings (SONAR) and from satellite altimetry. The bathymetric component of SRTM30_PLUS gets regularly updated with new or improved data sets in order to continuously improve and refine the description of the sea floor geometry. Over land areas, SRTM30 data from the USGS is included. SRTM30_PLUS provides background information for Google Earth and Google Maps.

ETOPO1 (2009)

The ETOPO1 1-arcmin global relief model, produced by the National Geophysical Data Center (Colorado), provides two layers of relief information. One layer represents the global relief including bedrock over Antarctica and Greenland, and another layer the global relief including ice surface heights. Both layers include bathymetry over the oceans and some of Earth's major lakes. ETOPO1 land topography and ocean bathymetry relies on SRTM30 topography and a multitude of bathymetric surveys that have been merged. Historic versions of ETOPO1 are the ETOPO2 and ETOPO5 relief models (2 and 5 arc-min resolution). The ETOPO1 global relief model is based on the 2001 Bedmap1 model of bedrock over Antarctica, which is now superseded by the significantly improved Bedmap2 bedrock data. The ETOPO1-contained information on ocean depths is superseded through several updates of the SRTM30_PLUS bathymetry.

Hypsometric Tints

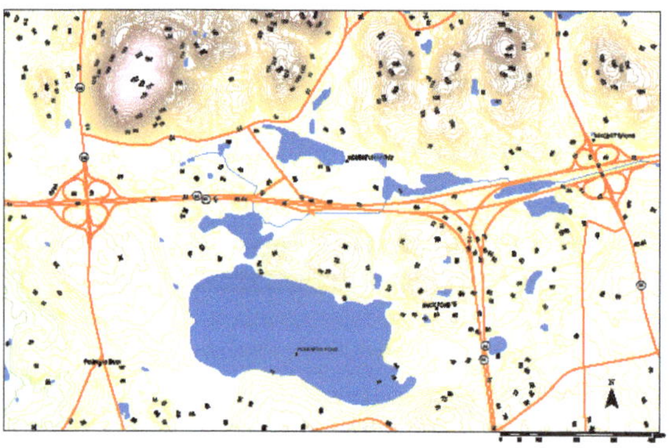

Hypsometric tinted contour lines

Hypsometric tints are colors used to indicate elevation, usually together with contour lines. They can be used to depict ranges of elevation as bands of color, usually in a graduated scheme, or as a color ramp applied to contour lines themselves. A typical scheme progresses from dark greens for lower elevations up through yellows/browns, and on to grays and white at the highest elevations. Hypsometric tinting of maps and globes is often accompanied by a similar method of bathymetric tinting to convey depth of oceans; lighter shades of blue represent shallower water such as the continental shelf and darker shades deeper regions.

History

In his map of central Italy, c.1503, Leonardo da Vinci introduced the cartographic convention of using color to indicate changes in elevation.

The Scottish map firm John Bartholomew and Son is credited with popularizing the technique, and its color scheme has become conventional: dark greens at low elevations, progressing through yellows and ochers, to browns and then grays and white at the highest elevations.

At right is "The very earliest rendition of a bathymetric map of an oceanic basin. Matthew Fontaine Maury published this map in 1853 in *Explanations and Sailing Directions to Accompany the Wind and Current Charts.*"

Uses

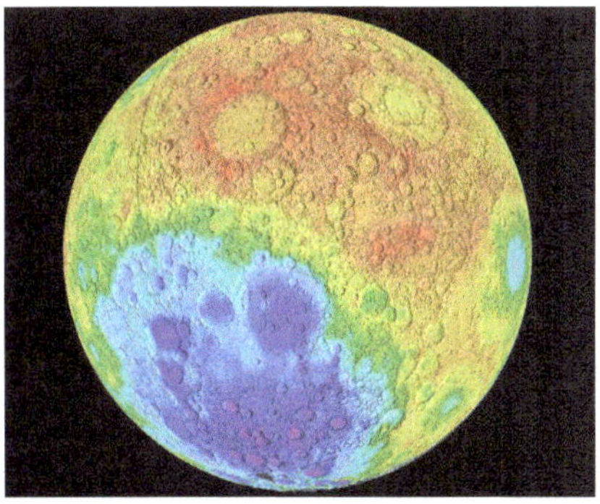

The Moon with hypsometric tints of red for the highest points and purple for the lowest.

Compare the information available in this image of a slightly different face of the Moon to the WorldWind depiction above.

Similar to false-color imagery, hypsometric tints can be used to make geographic information more accessible, as with this image of lunar topography.

Some cartographers have suggested that hypsometric tints are often used as decoration, rather than for informational purposes:

...the current popularity of hypsometric tints has more to do with production ease and pretty colors than it does with our interest in elevation. ... With hypsometric tints, the end result is often a map with pleasing colors that blend softly into one another in an orderly fashion, a design trait that people find attractive, even if they don't necessarily know or care about elevations.

Geovisualization

Geovisualization or Geovisualisation, short for Geographic Visualization, refers to a set of tools and techniques supporting the analysis of geospatial data through the use of interactive visualization.

Like the related fields of scientific visualization and information visualization geovisualization emphasizes knowledge construction over knowledge storage or information transmission. To do this, geovisualization communicates geospatial information in ways that, when combined with human understanding, allow for data exploration and decision-making processes.

Traditional, static maps have a limited exploratory capability; the graphical representations are inextricably linked to the geographical information beneath. GIS and geovisualization allow for more interactive maps; including the ability to explore different layers of the map, to zoom in or out, and to change the visual appearance of the map, usually on a computer display. Geovisualization represents a set of cartographic technologies and practices that take advantage of the ability of modern microprocessors to render changes to a map in real time, allowing users to adjust the mapped data on the fly.

History

The term visualization is first mentioned in the cartographic literature at least as early as 1953, in an article by University of Chicago geographer Allen K. Philbrick. New developments in the field of computer science prompted the National Science Foundation to redefine the term in a 1987 report which placed visualization at the convergence of computer graphics, image processing, computer vision, computer-aided design, signal processing, and user interface studies and emphasized both the knowledge creation and hypothesis generation aspects of scientific visualization.

Geovisualization developed as a field of research in the early 1980s, based largely on the work of French graphic theorist Jacques Bertin. Bertin's work on cartographic design and information visualization share with the National Science Foundation report a focus on the potential for the use of "dynamic visual displays as prompts for scientific insight and on the methods through which dynamic visual displays might leverage perceptual cognitive processes to facilitate scientific thinking".

Geovisualization has continued to grow as a subject of practice and research. The International Cartographic Association (ICA) established a Commission on Visualization & Virtual Environments in 1995.

Related Fields

Geovisualization is closely related to other visualization fields, such as scientific visualization and information visualization. Owing to its roots in cartography, geovisualization contributes to these other fields by way of the map metaphor, which "has been widely used to visualize non-geographic information in the domains of information visualization and domain knowledge visualization." It is also related to urban simulation.

Practical Applications

Geovisualization has made inroads in a diverse set of real-world situations calling for the decision-making and knowledge creation processes it can provide. The following list provides a summary of some of these applications as they are discussed in the geovisualization literature. E

Wildland Fire Fighting

Firefighters have been using sandbox environments to rapidly and physically model topography and fire for wildfire incident command strategic planning. The SimTable is a 3D interactive fire simulator, bringing sandtable exercises to life. The SimTable uses advanced computer simulations to model fires in any area, including local neighborhoods, utilizing actual slope, terrain, wind speed/direction, vegetation, and other factors. SimTable Models were used in Arizona's largest fire on record, the Wallow Fire.

Forestry

Geovisualizers, working with European foresters, used CommonGIS and Visualization Toolkit (VTK) to visualize a large set of spatio-temporal data related to European forests, allowing the data to be explored by non-experts over the Internet. The report summarizing this effort "uncovers a range of fundamental issues relevant to the broad field of geovisualization and information visualization research".

The research team cited the two major problems as the inability of the geovisualizers to convince the foresters of the efficacy of geovisualization in their work and the foresters' misgivings over the dataset's accessibility to non-experts engaging in "uncontrolled exploration". While the geovisualizers focused on the ability of geovisualization to aid in knowledge construction, the foresters preferred the information-communication role of more traditional forms of cartographic representation.

Archaeology

Geovisualization provides archaeologists with a potential technique for mapping unearthed archaeological environments as well as for accessing and exploring archaeological data in three dimensions.

The implications of geovisualization for archaeology are not limited to advances in archaeological theory and exploration but also include the development of new, collaborative relationships between archaeologists and computer scientists.

Environmental Studies

Geovisualization tools provide multiple stakeholders with the ability to make balanced environmental decisions by taking into account the "the complex interacting factors that should be taken into account when studying environmental changes". Geovisualization users can use a georeferenced model to explore a complex set of environmental data, interrogating a number of scenarios or policy options to determine a best fit.

Urban Planning

Both planners and the general public can use geovisualization to explore real-world environments and model 'what if' scenarios based on spatio-temporal data. While geovisualization in the preceding fields may be divided into two separate domains—the private domain, in which professionals use geovisualization to explore data and generate hypotheses, and the public domain, in which these professionals present their "visual thinking" to the general public—planning relies more heavily than many other fields on collaboration between the general public and professionals.

Planners use geovisualization as a tool for modeling the environmental interests and policy concerns of the general public. Jiang et al. mention two examples, in which "3D photo-realistic representations are used to show urban redevelopment [and] dynamic computer simulations are used to show possible pollution diffusion over the next few years." The widespread use of the Internet by the general public has implications for these collaborative planning efforts, leading to increased participation by the public while decreasing the amount of time it takes to debate more controversial planning decisions.

References

- Johnson, Anthony, Solving Stonehenge: The New Key to an Ancient Enigma. (Thames & Hudson, 2008) ISBN 978-0-500-05155-9
- Hong-Sen Yan & Marco Ceccarelli (2009), International Symposium on History of Machines and

Mechanisms: Proceedings of HMM 2008, Springer, p. 107, ISBN 1-4020-9484-1

- Lewis, M. J. T. (2001-04-23). Surveying Instruments of Greece and Rome. Cambridge University Press. ISBN 9780521792974. Retrieved 30 August 2012.
- National Cooperative Highway Research Program: Collecting, Processing and Integrating GPS data into GIS, p. 40. Published by Transportation Research Board, 2002 ISBN 0-309-06916-5, ISBN 978-0-309-06916-8
- Ronalds, B.F. (2016). Sir Francis Ronalds: Father of the Electric Telegraph. London: Imperial College Press. ISBN 978-1-78326-917-4.
- Schowengerdt, Robert A. (2007). Remote sensing: models and methods for image processing (3rd ed.). Academic Press. p. 2. ISBN 978-0-12-369407-2.
- Schott, John Robert (2007). Remote sensing: the image chain approach (2nd ed.). Oxford University Press. p. 1. ISBN 978-0-19-517817-3.
- Liu, Jian Guo & Mason, Philippa J. (2009). Essential Image Processing for GIS and Remote Sensing. Wiley-Blackwell. p. 4. ISBN 978-0-470-51032-2.
- Dawson, Alan (1997). The Hewitts and Marilyns of England. Glasgow: TACit Press. There are several related booklets covering Britain and Ireland. ISBN 0-9522680-7-8.
- Hirt, C. "Digital Terrain Models.". Encyclopedia of Geodesy: 1–6. doi:10.1007/978-3-319-02370-0_31-1. ISBN 978-3-319-01868-3. Retrieved February 10, 2016.
- Huerta, Michael P. (18 June 2014). "Interpretation of the Special Rule for Model Aircraft" (PDF). FAA. Retrieved 5 October 2016.
- "U.S. Transportation Secretary Foxx Announces FAA Exemptions for Commercial UAS Movie and TV Production". Press Release. FAA. 25 September 2014. Retrieved 4 October 2014.
- Cheves, Marc. "Geodimeter-The First Name in EDM". http://www.profsurv.com/magazine/. Archived from the original on March 10, 2014. Retrieved 2014-07-20.

5

Cartography: An Overview

Cartography is the study of creating and applying maps. The types of maps mentioned in this section are raised-relief maps, pictorial maps, fantasy maps, city maps and geologic maps. The other aspects of cartography examined are cartographic labeling, cartographic generalization, map projection and contour line. This section is an overview of the subject matter incorporating all the major aspects of cartography.

Cartography

A medieval depiction of the Ecumene (1482, Johannes Schnitzer, engraver), constructed after the coordinates in Ptolemy's Geography and using his second map projection. The translation into Latin and dissemination of *Geography* in Europe, in the beginning of the 15th century, marked the rebirth of scientific cartography, after more than a millennium of stagnation.

Cartography (from Greek χάρτης *khartēs*, "papyrus, sheet of paper, map"; and γράφειν *graphein*, "write") is the study and practice of making maps. Combining science, aesthetics, and technique, cartography builds on the premise that reality can be modeled in ways that communicate spatial information effectively.

The fundamental problems of traditional cartography are to:

- Set the map's agenda and select traits of the object to be mapped. This is the concern of map editing. Traits may be physical, such as roads or land masses, or may be abstract, such as toponyms or political boundaries.

- Represent the terrain of the mapped object on flat media. This is the concern of map projections.
- Eliminate characteristics of the mapped object that are not relevant to the map's purpose. This is the concern of generalization.
- Reduce the complexity of the characteristics that will be mapped. This is also the concern of generalization.
- Orchestrate the elements of the map to best convey its message to its audience. This is the concern of map design.

Modern cartography constitutes many theoretical and practical foundations of geographic information systems.

History

The earliest known map is a matter of some debate, both because the definition of "map" is not sharp and because some artifacts speculated to be maps might actually be something else. A wall painting, which may depict the ancient Anatolian city of Çatalhöyük (previously known as Catal Huyuk or Çatal Hüyük), has been dated to the late 7th millennium BCE. Among the prehistoric alpine rock carvings of Mount Bego (F) and Valcamonica (I), dated to the 4th millennium BCE, geometric patterns consisting of dotted rectangles and lines are widely interpreted in archaeological literature as a plan depiction of cultivated plots. Other known maps of the ancient world include the Minoan "House of the Admiral" wall painting from c. 1600 BCE, showing a seaside community in an oblique perspective and an engraved map of the holy Babylonian city of Nippur, from the Kassite period (14th – 12th centuries BCE). The oldest surviving world maps are the Babylonian world maps from the 9th century BCE. One shows Babylon on the Euphrates, surrounded by a circular landmass showing Assyria, Urartu and several cities, in turn surrounded by a "bitter river" (Oceanus), with seven islands arranged around it. Another depicts Babylon as being further north from the center of the world.

Valcamonica rock art (I), Paspardo r. 29, topographic composition, 4th millennium BC

The *Bedolina Map* and its tracing, 6th–4th century BC

Copy (1472) of St. Isidore's TO map of the world.

The ancient Greeks and Romans created maps, beginning at latest with Anaximander in the 6th century BC. In the 2nd century AD, Ptolemy produced his treatise on cartography, Geographia. This contained Ptolemy's world map – the world then known to Western society *(Ecumene).* As early as the 8th century, Arab scholars were translating the works of the Greek geographers into Arabic.

In ancient China, geographical literature spans back to the 5th century BC. The oldest extant Chinese maps come from the State of Qin, dated back to the 4th century BC, during the Warring States period. In the book of the *Xin Yi Xiang Fa Yao*, published in 1092 by the Chinese scientist Su Song, a star map on the equidistant cylindrical projection. Although this method of charting seems to have existed in China even prior to this publication and scientist, the greatest significance of the star maps by Su Song is that they represent the oldest existent star maps in printed form.

Early forms of cartography of India included the locations of the Pole star and other constellations of use. These charts may have been in use by the beginning of the Common Era for purposes of navigation.

Mappa mundi are the Medieval European maps of the world. Approximately 1,100 mappae mundi are known to have survived from the Middle Ages. Of these, some 900 are found illustrating manuscripts and the remainder exist as stand-alone documents.

The *Tabula Rogeriana*, drawn by Muhammad al-Idrisi for Roger II of Sicily in 1154

The Arab geographer Muhammad al-Idrisi produced his medieval atlas *Tabula Rogeriana* in 1154. He incorporated the knowledge of Africa, the Indian Ocean and the Far East, gathered by Arab merchants and explorers with the information inherited from the classical geographers to create the most accurate map of the world up until his time. It remained the most accurate world map for the next three centuries.

Europa regina in Sebastian Münster's "*Cosmographia*", 1570

In the Age of Exploration, from the 15th century to the 17th century, European cartographers both copied earlier maps (some of which had been passed down for centuries) and drew their own based on explorers' observations and new surveying techniques. The invention of the magnetic compass, telescope and sextant enabled increasing accuracy. In 1492, Martin Behaim, a German cartographer, made the oldest extant globe of the Earth.

Johannes Werner refined and promoted the Werner projection. In 1507, Martin Waldseemüller produced a globular world map and a large 12-panel world wall map (*Universalis Cosmographia*) bearing the first use of the name "America". Portuguese cartographer Diego Ribero was the author of the first known planisphere with a graduated Equator (1527). Italian cartographer Battista Agnese produced at least 71 manuscript atlases of sea charts.

Due to the sheer physical difficulties inherent in cartography, map-makers frequently lifted material from earlier works without giving credit to the original cartographer. For example, one of the most famous early maps of North America is unofficially known as the "Beaver Map", published in 1715 by Herman Moll. This map is an exact reproduction of a 1698 work by Nicolas de Fer. De Fer in turn had copied images that were first printed in books by Louis Hennepin, published in 1697, and François Du Creux, in 1664. By the 18th century, map-makers started to give credit to the original engraver by printing the phrase "After [the original cartographer]" on the work.

Technological Changes

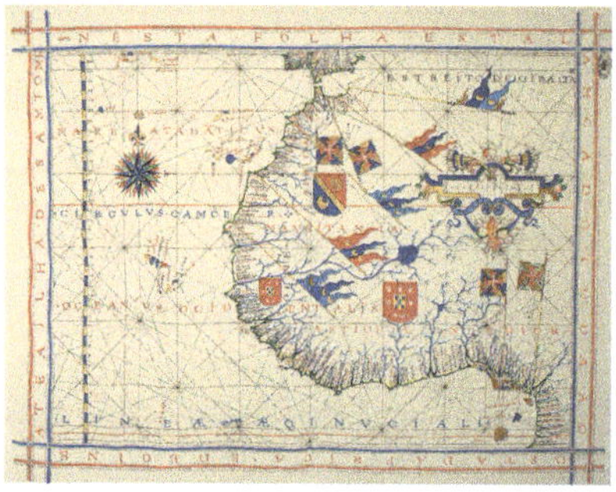

A pre-Mercator nautical chart of 1571, from Portuguese cartographer Fernão Vaz Dourado (c. 1520–c. 1580). It belongs to the so-called *plane chart* model, where observed latitudes and magnetic directions are plotted directly into the plane, with a constant scale, as if the Earth were a plane (Portuguese National Archives of Torre do Tombo, Lisbon).

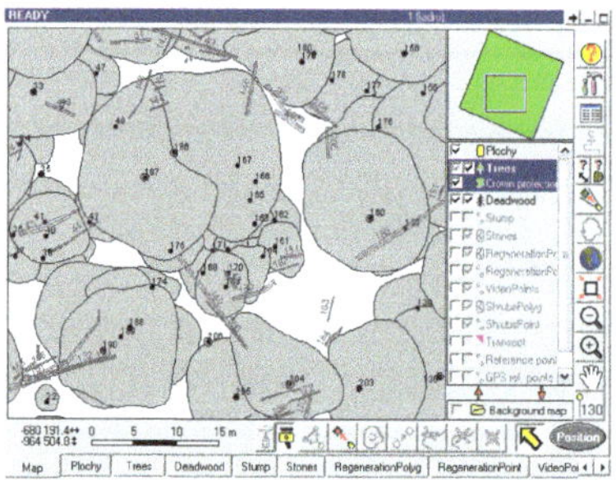

Mapping can be done with GPS and laser rangefinder directly in the field. Image shows mapping of forest structure (position of trees, dead wood and canopy).

In cartography, technology has continually changed in order to meet the demands of new generations of mapmakers and map users. The first maps were manually constructed with brushes and parchment; therefore, varied in quality and were limited in distribution. The advent of magnetic devices, such as the compass and much later, magnetic storage devices, allowed for the creation of far more accurate maps and the ability to store and manipulate them digitally.

Advances in mechanical devices such as the printing press, quadrant and vernier, allowed for the mass production of maps and the ability to make accurate reproductions from more accurate data. Optical technology, such as the telescope, sextant and other devices that use telescopes, allowed for accurate surveying of land and the ability of mapmakers and navigators to find their latitude by measuring angles to the North Star at night or the sun at noon.

Advances in photochemical technology, such as the lithographic and photochemical processes, have allowed for the creation of maps that have fine details, do not distort in shape and resist moisture and wear. This also eliminated the need for engraving, which further shortened the time it takes to make and reproduce maps.

In the 20th century, Aerial photography, satellite imagery, and remote sensing provided efficient, precise methods for mapping physical features, such as coastlines, roads, buildings, watersheds, and topography. Advancements in electronic technology ushered in another revolution in cartography. Ready availability of computers and peripherals such as monitors, plotters, printers, scanners (remote and document) and analytic stereo plotters, along with computer programs for visualization, image processing, spatial analysis, and database management, democratized and greatly expanded the making of maps. The ability to superimpose spatially located variables onto existing maps created new uses for maps and new industries to explore and exploit these potentials.

These days most commercial-quality maps are made using software that falls into one of three main types: CAD, GIS and specialized illustration software. Spatial information can be stored in a database, from which it can be extracted on demand. These tools lead to increasingly dynamic, interactive maps that can be manipulated digitally.

With the field rugged computers, GPS and laser rangefinders, it is possible to perform mapping directly in the terrain.

Deconstruction

There are technical and cultural aspects to the producing maps. In this sense, maps are biased. The study of bias, influence, and agenda in making a map is what comprise a map's deconstruction. A central tenet of deconstructionism is that maps have power. Other assertions are that maps are inherently biased and that we search for metaphor and rhetoric in maps.

It was the Europeans who promoted an epistemological understanding of the map as early as the 17th century. An example of this understanding is that, "[European reproduction of terrain on maps] reality can be expressed in mathematical terms; that systematic observation and measurement offer the only route to cartographic truth...". 17th century map-makers were careful and precise in their strategic approaches to maps based on a scientific model of knowledge. Popular belief at the time was that this scientific approach to cartography was immune to the social atmosphere.

A common belief is that science heads in a direction of progress, and thus leads to more accurate representations of maps. In this belief European maps must be superior to others, which necessarily employed different map-making skills. "There was a 'not cartography' land where lurked an army of inaccurate, heretical, subjective, valuative, and ideologically distorted images. Cartographers developed a 'sense of the other' in relation to nonconforming maps."

Though cartography has been a target of much criticism in recent decades, a cartographer's 'black box' always seemed to be naturally defended to the point where it overcame the criticism. However, to later scholars in the field, it was evident that cultural influences dominate map-making. For instance, certain abstracts on maps and the map-making society itself describe the social influences on the production of maps. This social play on cartographic knowledge "...produces the 'order' of [maps'] features and the 'hierarchies of its practices.'"

Depictions of Africa are a common target of deconstructionism. According to deconstructionist models, cartography was used for strategic purposes associated with imperialism and as instruments and representations of power during the conquest of Africa. The depiction of Africa and the low latitudes in general on the Mercator projection has been interpreted as imperialistic and as symbolic of subjugation due to the diminished proportions of those regions compared to higher latitudes where the European powers were concentrated.

Maps furthered imperialism and colonization of Africa through practical ways such as showing basic information like roads, terrain, natural resources, settlements, and communities. Through this, maps made European commerce in Africa possible by showing potential commercial routes, and made natural resource extraction possible by depicting locations of resources. Such maps also enabled military conquests and made them more efficient, and imperial nations further used them to put their conquests on display. These same maps were then used to cement territorial claims, such as at the Berlin Conference of 1884–1885.

Before 1749, maps of the African continent had African kingdoms drawn with assumed or contrived boundaries, with unknown or unexplored areas having drawings of animals, imaginary physical geographic features, and descriptive texts. In 1748 Jean B. B. d'Anville created the first map of the African continent that had blank spaces to represent the unknown territory. This was revolutionary in cartography and the representation of power associated with map making.

Map Types

General vs. Thematic Cartography

In understanding basic maps, the field of cartography can be divided into two general categories: general cartography and thematic cartography. General cartography involves those maps that are constructed for a general audience and thus contain a variety of features. General maps exhibit many reference and location systems and often are produced in a series. For example, the 1:24,000 scale topographic maps of the United States Geological Survey (USGS) are a standard as compared to the 1:50,000 scale Canadian maps. The government of the UK produces the classic 1:50,000 (replacing the older 1 inch to 1 mile) "Ordnance Survey" maps of the entire UK and with a range of correlated larger- and smaller-scale maps of great detail. Many private mapping companies have also produce thematic map series.

Small section of an orienteering map.

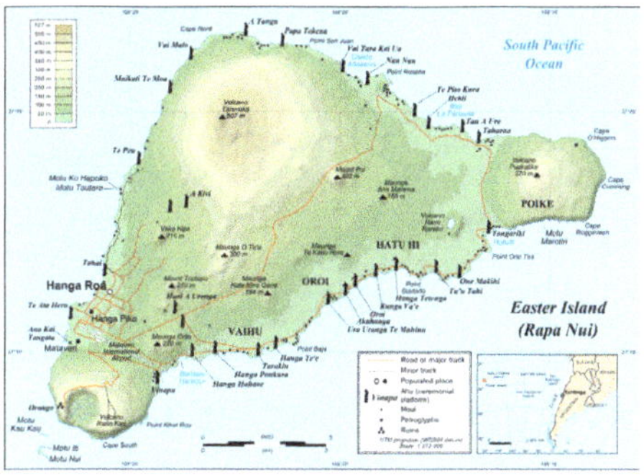

Topographic map of Easter Island.

Thematic cartography involves maps of specific geographic themes, oriented toward specific audiences. A couple of examples might be a dot map showing corn production in Indiana or a shaded area map of Ohio counties, divided into numerical choropleth classes. As the volume of geographic data has exploded over the last century, thematic cartography has become increasingly useful and necessary to interpret spatial, cultural and social data.

An orienteering map combines both general and thematic cartography, designed for a very specific user community. The most prominent thematic element is shading, that indicates degrees of difficulty of travel due to vegetation. The vegetation itself is not identified, merely classified by the difficulty ("fight") that it presents.

Topographic vs. Topological

A topographic map is primarily concerned with the topographic description of a place, including (especially in the 20th and 21st centuries) the use of contour lines showing eleva-tion. Terrain or relief can be shown in a variety of ways.

A topological map is a very general type of map, the kind one might sketch on a napkin. It often disregards scale and detail in the interest of clarity of communicating specific route or relational information. Beck's London Underground map is an iconic example. Though the most widely used map of "The Tube," it preserves little of reality: it varies scale constantly and abruptly, it straightens curved tracks, and it contorts directions. The only topography on it is the River Thames, letting the reader know whether a station is north or south of the river. That and the topology of station order and interchanges between train lines are all that is left of the geographic space. Yet those are all a typical passenger wishes to know, so the map fulfils its purpose.

Map Design

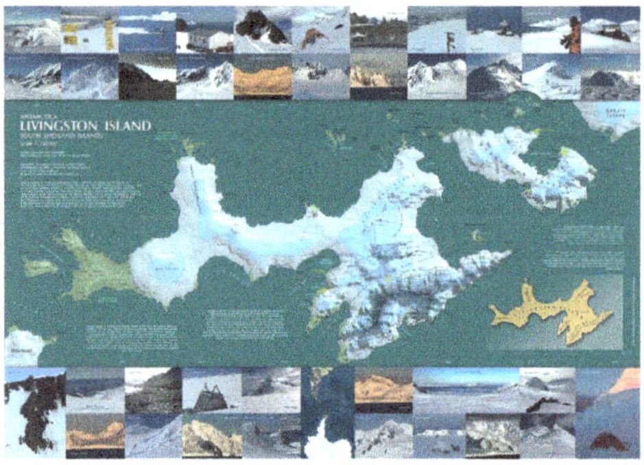

Illustrated map.

Map Purpose and Selection of Information

Arthur H. Robinson, an American cartographer influential in thematic cartography, stated that a map not properly designed "will be a cartographic failure." He also claimed, when considering all aspects of cartography, that "map design is perhaps the most complex." Robinson codified the mapmaker's understanding that a map must be designed foremost with consideration to the audience and its needs.

From the very beginning of mapmaking, maps "have been made for some particular purpose or set of purposes". The intent of the map should be illustrated in a manner in which the percipient acknowledges its purpose in a timely fashion. The term *percipient* refers to the person receiving information and was coined by Robinson. The principle of figure-ground refers to this notion of engaging the user by presenting a clear presentation, leaving no confusion concerning the purpose of the map. This will enhance the user's experience and keep his attention. If the user is unable to identify what is being demonstrated in a reasonable fashion, the map may be regarded as useless.

Making a meaningful map is the ultimate goal. Alan MacEachren explains that a well designed map "is convincing because it implies authenticity" (1994, pp. 9). An interesting map will no doubt engage a reader. Information richness or a map that is multivariate shows relationships within the map. Showing several variables allows comparison, which adds to the meaningfulness of the map. This also generates hypothesis and stimulates ideas and perhaps further research. In order to convey the message of the map, the creator must design it in a manner which will aid the reader in the overall understanding of its purpose. The title of a map may provide the "needed link" necessary for communicating that message, but the overall design of the map fosters the manner in which the reader interprets it (Monmonier, 1993, pp. 93).

In the 21st century it is possible to find a map of virtually anything from the inner workings of the human body to the virtual worlds of cyberspace. Therefore, there are now a huge variety of different styles and types of map – for example, one area which has evolved a specific and recognisable variation are those used by public transport organisations to guide passengers, namely urban rail and metro maps, many of which are loosely based on 45 degree angles as originally perfected by Harry Beck and George Dow.

Naming Conventions

Most maps use text to label places and for such things as the map title, legend and other information. Although maps are often made in one specific language, place names often differ between languages. So a map made in English may use the name *Germany* for that country, while a German map would use *Deutschland* and a French map *Allemagne*. A non-native term for a place is referred to as an exonym.

In some cases the correct name is not clear. For example, the nation of Burma officially changed its name to Myanmar, but many nations do not recognize the ruling junta and continue to use *Burma*. Sometimes an official name change is resisted in other languages and the older name may remain in common use. Examples include the use of *Saigon* for Ho Chi Minh City, *Bangkok* for Krung Thep and *Ivory Coast* for Côte d'Ivoire.

Difficulties arise when transliteration or transcription between writing systems is required. Some well-known places have well-established names in other languages and writing systems, such as *Russia* or *Rußland* for Росси́я, but in other cases a system of transliteration or transcription is required. Even in the former case, the exclusive use of an exonym may be unhelpful for the map user. It will not be much use for an English user of a map of Italy to show Livorno *only* as "Leghorn" when road signs and railway timetables show it as "Livorno". In transliteration, the characters in one script are represented by characters in another. For example, the Cyrillic letter *P* is usually written as *R* in the Latin script, although in many cases it is not as simple as a one-for-one equivalence. Systems exist for transliteration of Arabic, but the results may vary. For example, the Yemeni city of Mocha is written variously in English as Mocha, Al Mukha, al-Mukhā, Mocca and Moka. Transliteration systems are based on relating written symbols to one another, while transcription is the attempt to spell in one language the phonetic sounds of another. Chinese writing is now usually converted to the Latin alphabet through the Pinyin phonetic transcription systems. Other systems were used in the past, such as Wade-Giles, resulting in the city being spelled *Beijing* on newer English maps and *Peking* on older ones.

Further difficulties arise when countries, especially former colonies, do not have a strong national geographic naming standard. In such cases, cartographers may have to choose between various phonetic spellings of local names versus older imposed, sometimes resented, colonial names. Some countries have multiple official languages, resulting in multiple official placenames. For example, the capital of Belgium is both *Brussel* and *Bruxelles*. In Canada, English and French are official languages and places have names in both languages. British Columbia is also officially named *la Colombie-Britannique*. English maps rarely show the French names outside of Quebec, which itself is spelled *Québec* in French.

The study of placenames is called toponymy, while that of the origin and historical usage of placenames as words is etymology.

In order to improve legibility or to aid the illiterate, some maps have been produced using pictograms to represent places. The iconic example of this practice is Lance Wyman's early plans for the Mexico City Metro, on which stations were shown simply as stylized logos. Wyman also prototyped such a map for the Washington Metro, though ultimately the idea was rejected. Other cities experimenting with such maps are Fukuoka, Guadalajara and Monterrey.

Map Symbology

Cartographic symbology encodes information on the map in ways intended to convey information to the map reader efficiently, taking into consideration the limited space on the map, models of human understanding through visual means, and the likely cultural background and education of the map reader. Symbology may be implicit, using universal elements of design, or may be more specific to cartography or even to the map.

A map may have any of many kinds of symbolization. Some examples are:

- A legend, or key, explains the map's pictorial language.
- A title indicates the region and perhaps the theme that the map portrays.
- A neatline frames the entire map image.
- A compass rose or north arrow provides orientation.
- An overview map gives global context for the primary map.
- A bar scale translates between map measurements and real distances.
- A map projection provides a way to represent the curved surface on the plane of the map.

The map may declare its sources, accuracy, publication date and authorship, and so forth. The map image itself portrays the region.

Map coloring is another form of symbology, one whose importance can reach beyond aesthetic. In complex thematic maps, for example, the color scheme's structure can critically affect the reader's ability to understand the map's information. Modern computer displays and print technologies can reproduce much of the gamut that humans can perceive, allowing for intricate exploitation of human visual discrimination in order to convey detailed information.

Quantitative symbols give a visual indication of the magnitude of the phenomenon that the symbol represents. Two major classes of symbols are used to portray quantity. Proportional symbols change size according to phenomenon's magnitude, making them appropriate for representing statistics. Choropleth maps portray data collection areas, such as counties or census tracts, with color. Using color this way, the darkness and intensity (or value) of the color is evaluated by the eye as a measure of intensity or concentration.

Map Key or Legend

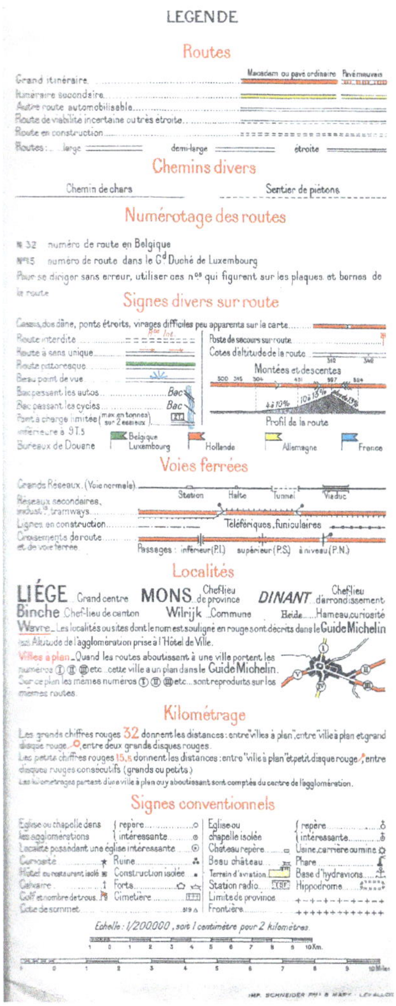

Legend or key of a French road map (Michelin 1940)

The map key, or legend, describes how to interpret the map's symbols and may give details of publication and authorship.

Examples of Point Symbols

Symbol	Explanation
	mine (Hammer and pick symbol), former mine
	castle, Burg
	church, chapel, monastery (♁)
	monument
	Hotel
	airport
	railway station
	Tourist information

Map Generalization

A good map has to compromise between portraying the items of interest (or themes) in the right place on the map, and the need to show that item using text or a symbol, which take up space on the map and might displace some other item of information. The cartographer is thus constantly making judgements about what to include, what to leave out and what to show in a *slightly* incorrect place. This issue assumes more importance as the scale of the map gets smaller (i.e. the map shows a larger area) because the information shown on the map takes up more space *on the ground*. A good example from the late 1980s was the Ordnance Survey's first digital maps, where the *absolute* positions of major roads were sometimes a scale distance of hundreds of metres away from ground truth, when shown on digital maps at scales of 1:250,000 and 1:625,000, because of the overriding need to annotate the features.

Map Projections

The Earth being spherical, any flat representation generates distortions such that shapes and areas cannot both be conserved simultaneously, and distances can never all be preserved. The mapmaker must choose a suitable *map projection* according to the space to be mapped and the purpose of the map.

Cartographic Errors

Some maps contain deliberate errors or distortions, either as propaganda or as a "watermark" to help the copyright owner identify infringement if the error appears in competitors' maps. The latter often come in the form of nonexistent, misnamed, or misspelled "trap streets". Other names and forms for this are paper townsites, fictitious entries, and copyright easter eggs.

Another motive for deliberate errors is cartographic "vandalism": a mapmaker wishing to leave his or her mark on the work. Mount Richard, for example, was a fictitious peak on the Rocky Mountains' continental divide that appeared on a Boulder County, Colorado map in the early 1970s. It is believed to be the work of draftsman Richard Ciacci. The fiction was not discovered until two years later.

Sandy Island (New Caledonia) is an example of a fictitious location that stubbornly survives, reappearing on new maps copied from older maps while being deleted from other new editions.

Map

A map is a symbolic depiction highlighting relationships between elements of some space, such as objects, regions, and themes.

World map (1689, Amsterdam)

Many maps are static two-dimensional, geometrically accurate (or approximately accurate) representations of three-dimensional space, while others are dynamic or interactive, even three-dimensional. Although most commonly used to depict geography, maps may represent any space, real or imagined, without regard to context or scale; e.g. brain mapping, DNA mapping and extraterrestrial mapping.

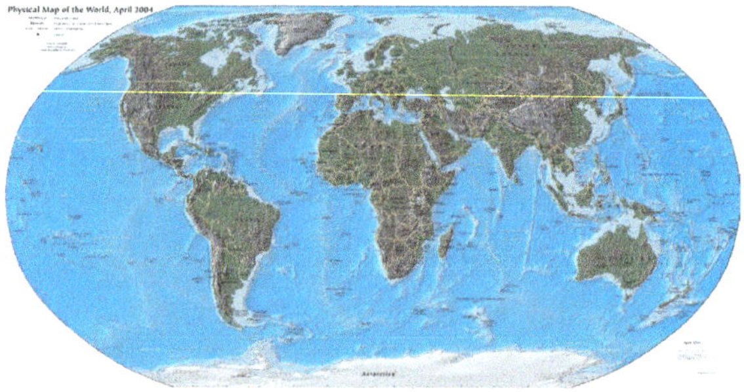

World map (2004, CIA World Factbook)

Although the earliest maps known are of the heavens, geographic maps of territory have a very long tradition and exist from ancient times. The word “map” comes from the medieval Latin *Mappa mundi*, wherein *mappa* meant napkin or cloth and *mundi* the world. Thus, “map” became the shortened term referring to a two-dimensional representation of the surface of the world.

Geographic Maps

A celestial map from the 17th century, by the cartographer Frederik de Wit

Cartography or *map-making* is the study and practice of crafting representations of the Earth upon a flat surface, and one who makes maps is called a cartographer.

Road maps are perhaps the most widely used maps today, and form a subset of navigational maps, which also include aeronautical and nautical charts, railroad network maps, and hiking and bicycling maps. In terms of quantity, the largest number of drawn map sheets is probably made up by local surveys, carried out by municipalities, utilities, tax assessors, emergency services providers, and other local agencies. Many national surveying projects have been carried out by the military, such as the British Ordnance Survey: a civilian government agency, internationally renowned for its comprehensively detailed work.

In addition to location information maps may also be used to portray contour lines indicating constant values of elevation, temperature, rainfall, etc.

Orientation of Maps

The Hereford Mappa Mundi from about 1300, Hereford Cathedral, England, is a classic "T-O" map with Jerusalem at centre, east toward the top, Europe the bottom left and Africa on the right.

The orientation of a map is the relationship between the directions on the map and the corresponding compass directions in reality. The word "orient" is derived from Latin *oriens*, meaning East. In the Middle Ages many maps, including the T and O maps, were drawn with East at the top (meaning that the direction "up" on the map corresponds to East on the compass). Today, the most common – but far from universal – cartographic convention is that North is at the top of a map. Several kinds of maps are often traditionally not oriented with North at the top:

- Maps from non-Western traditions are oriented a variety of ways. Old maps of Edo show the Japanese imperial palace as the "top", but also at the centre, of the map. Labels on the map are oriented in such a way that you cannot read them properly unless you put the imperial palace above your head.

- Medieval European T and O maps such as the Hereford Mappa Mundi were centred on Jerusalem with East at the top. Indeed, prior to the reintroduction of Ptolemy's *Geography* to Europe around 1400, there was no single convention in the West. Portolan charts, for example, are oriented to the shores they describe.

- Maps of cities bordering a sea are often conventionally oriented with the sea at the top.

- Route and channel maps have traditionally been oriented to the road or waterway they describe.

- Polar maps of the Arctic or Antarctic regions are conventionally centred on the pole; the direction North would be towards or away from the centre of the map, respectively. Typical maps of the Arctic have 0° meridian towards the bottom of the page; maps of the Antarctic have the 0° meridian towards the top of the page.

- Reversed maps, also known as *Upside-Down maps* or *South-Up maps*, reverse the *North is up* convention and have south at the top.

- Buckminster Fuller's Dymaxion maps are based on a projection of the Earth's sphere onto an icosahedron. The resulting triangular pieces may be arranged in any order or orientation.

- Modern digital GIS maps such as ArcMap typically project north at the top of the map, but use math degrees (0 is east, degrees increase counter-clockwise), rather than compass degrees (0 is north, degrees increase clockwise) for orientation of transects. Compass decimal degrees can be converted to math degrees by subtracting them from 450; if the answer is greater than 360, subtract 360.

Scale and accuracy

A 'global view map' of Europe, Western Asia and Africa.

Many, but not all, maps are drawn to a scale, expressed as a ratio such as 1:10,000, meaning that 1 of any unit of measurement on the map corresponds exactly, to 10,000 of that same unit on the ground. The scale statement may be taken as exact when the region mapped is small enough for the curvature of the Earth to be neglected, for example in a town planner's city map. Over larger regions where the curvature cannot be ignored we must use map projections from the curved surface of the Earth (sphere or ellipsoid) to the plane. The impossibility of flattening the sphere to the plane implies that no map projection can have constant scale: on most projections the best we can achieve is accurate scale on one or two lines (not necessarily straight) on the projection. Thus for map projections we must introduce the concept of point scale, which is a function of position, and strive to keep its variation within narrow bounds. Although the scale statement is nominal it is usually accurate enough for all but the most precise of measurements.

Large scale maps, say 1:10,000, cover relatively small regions in great detail and small scale maps, say 1:10,000,000, cover large regions such as nations, continents and the whole globe. The large/small terminology arose from the practice of writing scales as numerical fractions: 1/10,000 is larger than 1/10,000,000. There is no exact dividing line between large and small but 1/100,000 might well be considered as a medium scale. Examples of large scale maps are the 1:25,000 maps produced for hikers; on the other hand maps intended for motorists at 1:250,000 or 1:1,000,000 are small scale.

It is important to recognize that even the most accurate maps sacrifice a certain amount of accuracy in scale to deliver a greater visual usefulness to its user. For example, the width of roads and small streams are exaggerated when they are too narrow to be shown on the map at true scale; that is, on a printed map they would be narrower than could be perceived by the naked eye. The same applies to computer maps where the smallest unit is the pixel. A narrow stream say must be shown to have the width of a pixel even if at the map scale it would be a small fraction of the pixel width.

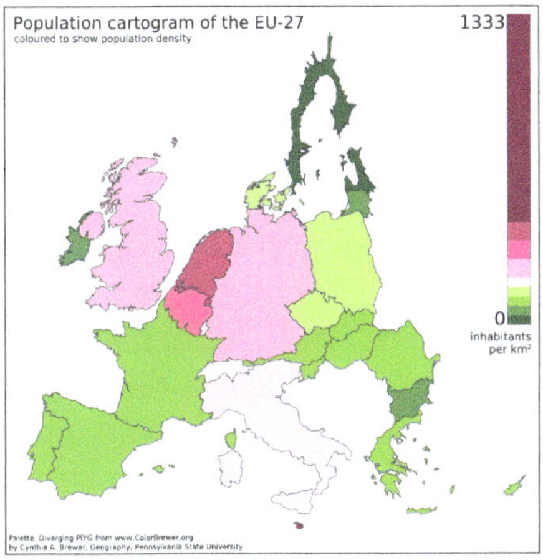

Cartogram: The EU distorted to show population distributions.

Some maps, called cartograms, have the scale deliberately distorted to reflect information other than land area or distance. For example, this map (at the right) of Europe has been distorted to show population distribution, while the rough shape of the continent is still discernible.

Another example of distorted scale is the famous London Underground map. The basic geographical structure is respected but the tube lines (and the River Thames) are smoothed to clarify the relationships between stations. Near the center of the map stations are spaced out more than near the edges of map.

Further inaccuracies may be deliberate. For example, cartographers may simply omit military installations or remove features solely in order to enhance the clarity of the map. For example, a road map may not show railroads, smaller waterways or other prominent non-road objects, and even if it does, it may show them less clearly (e.g. dashed or dotted lines/outlines) than the main roads. Known as decluttering, the practice makes the subject matter that the user is interested in easier to read, usually without sacrificing overall accuracy. Software-based maps often allow the user to toggle decluttering between ON, OFF and AUTO as needed. In AUTO the degree of decluttering is adjusted as the user changes the scale being displayed.

Map Types and Projections

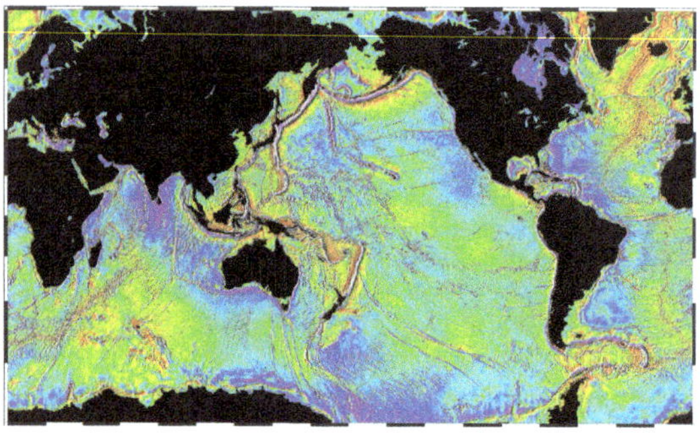

Map of large underwater features. (1995, NOAA)

Maps of the world or large areas are often either 'political' or 'physical'. The most important purpose of the political map is to show territorial borders; the purpose of the physical is to show features of geography such as mountains, soil type or land use including infrastructure such as roads, railroads and buildings. Topographic maps show elevations and relief with contour lines or shading. Geological maps show not only the physical surface, but characteristics of the underlying rock, fault lines, and subsurface structures. Maps that depict the surface of the Earth also use a projection, a way of translating the three-dimensional real surface of the geoid to a two-dimensional picture. Perhaps the best-known world-map projection is the Mercator projection, originally designed as a form of nautical chart. Aeroplane pilots use aeronautical charts based on a Lambert conformal conic projection, in which a cone is laid over the section of the earth to be mapped. The cone intersects the sphere (the earth) at one or two parallels which are chosen as standard lines. This allows the pilots to plot a great-circle route approximation on a flat, two-dimensional chart.

- Azimuthal or Gnomonic map projections are often used in planning air routes due to their ability to represent great circles as straight lines.
- Richard Edes Harrison produced a striking series of maps during and after World War II for Fortune magazine. These used "bird's eye" projections to emphasise globally strategic "fronts" in the air age, pointing out proximities and barriers not apparent on a conventional rectangular projection of the world.

Electronic Maps

From the last quarter of the 20th century, the indispensable tool of the cartographer has been the computer. Much of cartography, especially at the data-gathering survey level, has been subsumed by Geographic Information Systems (GIS). The functionality of maps has been greatly advanced by technology simplifying the superimposition of spatially located variables onto existing geographical maps. Having local information such as rainfall level, distribution of wildlife, or demographic data integrated within the map allows more efficient analysis and better decision making. In the pre-electronic age such superimposition of data led Dr. John Snow to identify the location of an outbreak of cholera. Today, it is used by agencies of the human kind, as diverse as wildlife conservationists and militaries around the world.

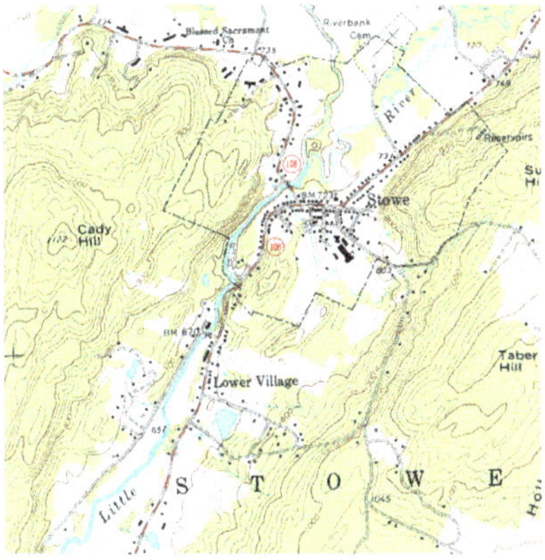

A USGS digital raster graphic.

Even when GIS is not involved, most cartographers now use a variety of computer graphics programs to generate new maps.

Interactive, computerised maps are commercially available, allowing users to *zoom in* or *zoom out* (respectively meaning to increase or decrease the scale), sometimes by replacing one map with another of different scale, centered where possible on the same point. In-car global navigation satellite systems are computerised maps with route-planning and advice facilities which monitor the user's position with the help of satellites. From the computer scientist's point of view, zooming in entails one or a combination of:

1. replacing the map by a more detailed one
2. enlarging the same map without enlarging the pixels, hence showing more detail by removing less information compared to the less detailed version
3. enlarging the same map with the pixels enlarged (replaced by rectangles of pixels); no additional detail is shown, but, depending on the quality of one's vision, possibly more detail can be seen; if a computer display does not show adjacent pixels really separate, but overlapping instead (this does not apply for an LCD, but may apply for a cathode ray tube), then replacing a pixel by a rectangle of pixels does show more detail. A variation of this method is interpolation.

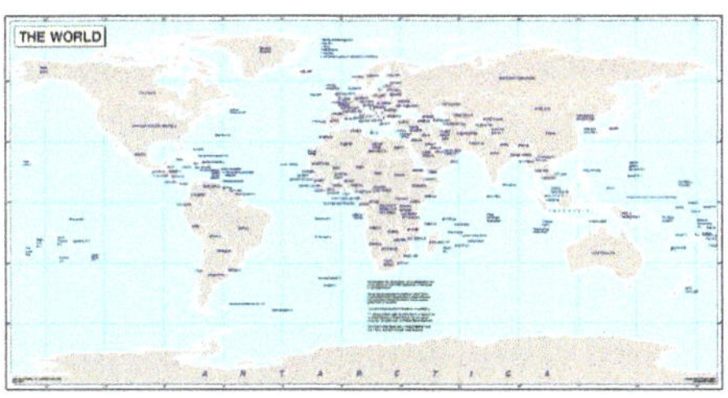

A world map in PDF format.

For example:

- Typically (2) applies to a Portable Document Format (PDF) file or other format based on vector graphics. The increase in detail is, of course, limited to the information contained in the file: enlargement of a curve may eventually result in a series of standard geometric figures such as straight lines, arcs of circles or splines.
- (2) may apply to text and (3) to the outline of a map feature such as a forest or building.
- (1) may apply to the text as needed (displaying labels for more features), while (2) applies to the rest of the image. Text is not necessarily enlarged when zooming in. Similarly, a road represented by a double line may or may not become wider when one zooms in.
- The map may also have layers which are partly raster graphics and partly vector graphics. For a single raster graphics image (2) applies until the pixels in the image file correspond to the pixels of the display, thereafter (3) applies.

Climatic Maps

The maps that reflect the territorial distribution of climatic conditions based on the results of long-term observations. Climatic maps can be compiled both for individual climatic features (temperature, precipitation, humidity) and for combinations of them at the earth's surface and in the upper layers of the atmosphere. Climatic maps afford a very convenient overview of the climatic features in a large region and permit values of climatic features to be compared in different parts of the region. Through interpolation the maps can be used to determine the values of climatic features in any particular spot.

Climatic maps generally apply to individual months and to the year as a whole, sometimes to the four seasons, to the growing period, and so forth. On maps compiled from the observations of ground meteorological stations, atmospheric pressure is converted to sea level. Air temperature maps are compiled both from the actual values observed on the surface of the earth and from values converted to sea level. The pressure field in free atmosphere is represented either by maps of the distribution of pressure at different standard altitudes—for example, at every kilometer above sea level—or by maps of baric topography on which altitudes (more precisely geopotentials) of the main isobaric surfaces (for example, 900, 800, and 700 millibars) counted off from sea level are plotted. The temperature, humidity, and wind on aeroclimatic maps may apply either to standard altitudes or to the main isobaric surfaces.

Isolines are drawn on maps of such climatic features as the long-term mean values (of atmospheric pressure, temperature, humidity, total precipitation, and so forth) to connect points with equal values of the feature in question—for example, isobars for pressure, isotherms for temperature, and isohyets for precipitation. Isoamplitudes are drawn on maps of amplitudes (for example, annual amplitudes of air temperature—that is, the differences between the mean temperatures of the warmest and coldest month). Isanomals are drawn on maps of anomalies (for example, deviations of the mean temperature of each place from the mean temperature of the entire latitudinal zone). Isolines of frequency are drawn on maps showing the frequency of a particular phenomenon (for example, annual number of days with a thunderstorm or snow cover). Isochrones are drawn on maps showing the dates of onset of a given phenomenon (for example, the first frost and appearance or disappearance of the snow cover) or the date of a particular value of a meteorological element in the course of a year (for example, passing of the mean daily air temperature through zero). Isolines of the mean numerical value of wind velocity or isotachs are drawn on wind maps (charts); the wind resultants and directions of prevailing winds are indicated by arrows of different length or arrows with different plumes; lines of flow are often drawn. Maps of the zonal and meridional components of wind are frequently compiled for the free atmosphere. Atmospheric pressure and wind are usually combined on climatic maps. Wind roses, curves showing the distribution of other meteorological elements, diagrams of the annual course of elements at individual stations, and the like are also plotted on climatic maps.

Maps of climatic regionalization, that is, division of the earth's surface into climatic zones and regions according to some classification of climates, are a special kind of climatic map.

Climatic maps are often incorporated into climatic atlases of varying geographic range (globe, hemispheres, continents, countries, oceans) or included in comprehensive atlases. Besides general climatic maps, applied climatic maps and atlases have great practical value. Aeroclimatic maps, aeroclimatic atlases, and agroclimatic maps are the most numerous.

Conventional Signs

The various features shown on a map are represented by conventional signs or symbols. For example, colors can be used to indicate a classification of roads. Those signs are usually explained in the margin of the map, or on a separately published characteristic sheet.

Some cartographers prefer to make the map cover practically the entire screen or sheet of paper, leaving no room "outside" the map for information about the map as a whole. These cartographers typically place such information in an otherwise "blank" region "inside" the map -- cartouche, map legend, title, compass rose, bar scale, etc. In particular, some maps contain smaller "sub-maps" in otherwise blank regions—often one at a much smaller scale showing the whole globe and where the whole map fits on that globe, and a few showing "regions of interest" at a larger scale in order to show details that wouldn't otherwise fit. Occasionally sub-maps use the same scale as the large map—a few maps of the contiguous United States include a sub-map to the same scale for each of the two non-contiguous states.

Labeling

To communicate spatial information effectively, features such as rivers, lakes, and cities need to be labeled. Over centuries cartographers have developed the art of placing names on even the densest of maps. Text placement or name placement can get mathematically very complex as the number of labels and map density increases. Therefore, text placement is time-consuming and labor-intensive, so cartographers and GIS users have developed automatic label placement to ease this process.

Non-geographical Spatial Maps

Maps exist of the Solar System, and other cosmological features such as star maps. In addition maps of other bodies such as the Moon and other planets are technically not *geo*graphical maps.

Non Spatial Maps

Diagrams such as schematic diagrams and Gantt charts and treemaps display logical relationships between items, and do not display spatial relationships at all.

Some maps, for example the London Underground map, are topological maps. Topological in nature, the distances are completely unimportant; only the connectivity is significant.

General-purpose Maps

General-purpose maps provide many types of information on one map. Most atlas maps, wall maps, and road maps fall into this category. The following are some features that might be shown on a general-purpose maps: bodies of water, roads, railway lines, parks, elevations, towns and cities, political boundaries, latitude and longitude, national and provincial parks. These maps give a broad understanding of location and features of an area. The reader may gain an understanding of the type of landscape, the location of urban places, and the location of major transportation routes all at once.

Types of Maps

- Atlas
- Climatic map
- Physical map

- Political map
- Street map
- Thematic map
- Weather map
- Relief map
- World map

Legal Regulation

Some countries required that all published maps represent their national claims regarding border disputes. For example:

- Within Russia, Google Maps shows Crimea as part of Russia.
- Both the Republic of India and the People's Republic of China require that all maps show areas subject to the Sino-Indian border dispute in their own favor.

In 2010, the People's Republic of China began requiring that all online maps served from within China be hosted there, making them subject to Chinese laws.

Types of Map

Raised-relief Map

Hand-made raised-relief map of the High Tatras in scale 1: 50 000

A raised-relief map or terrain model is a three-dimensional representation, usually of terrain, materialized as a physical artifact. When representing terrain, the vertical dimension is usually exaggerated by a factor between five and ten; this facilitates the visual recognition of terrain features.

History

If the account of Sima Qian (c. 145–86 BCE) in his *Records of the Grand Historian* is proven correct upon the unearthing of Qin Shi Huang's tomb, the raised-relief map has existed since the Qin dynasty (221–206 BCE) of China. Joseph Needham suggests that certain pottery vessels of the Han dynasty (202 BCE – 220 CE) showing artificial mountains as lid decorations may have influenced the raised-relief map.

The Han dynasty general Ma Yuan made a raised-relief map of valleys and mountains in a rice-constructed model of 32 CE. Such rice models were expounded on by the Tang dynasty (618–907) author Jiang Fang in his *Essay on the Art of Constructing Mountains with Rice* (c. 845). A raised-relief map made of wood representing all the provinces of the empire and put together like a giant 0.93 m^2 (10 ft^2) jigsaw puzzle was invented by Xie Zhuang (421–466) during the Liu Song dynasty (420–479).

Shen Kuo (1031-1095) created a raised-relief map using sawdust, wood, beeswax, and wheat paste. His wooden model pleased Emperor Shenzong of Song, who later ordered that all the prefects administering the frontier regions should prepare similar wooden maps which could be sent to the capital and stored in an archive.

In 1130, Huang Shang made a wooden raised-relief map which later caught the attention of the Neo-Confucian philosopher Zhu Xi, who tried to acquire it but instead made his own map out of sticky clay and wood. The map, made of eight pieces of wood connected by hinges, could be folded up and carried around by one person.

Later, Ibn Battuta (1304–1377) described a raised-relief map while visiting Gibraltar

In his 1665 paper for the *Philosophical Transactions of the Royal Society*, John Evelyn (1620–1706) believed that wax models imitating nature and bas relief maps were something entirely new from France. Some later scholars attributed the first raised-relief map to one Paul Dox, who represented the area of Kufstein in his raised-relief map of 1510.

Construction

Starting with a topographic map, one can cut out successive layers from some sheet material, with edges following the contour lines on the map. These may be assembled in a stack to obtain a rough approximation of the terrain. This method is commonly used as the base for architectural models, and is usually done without vertical exaggeration. For models of landforms, the stack can then be smoothed by filling with some material. This model may be used directly, or for greater durability a mold may be made from it. This mold may then be used to produce a plaster model.

A combination of computer numerical control (CNC) machining a master model, and vacuum forming SAM copies from this, can be been used to rapidly mass-produce raised-relief maps.

Another method which is becoming more widespread is the use of 3D printing. With the rapid development of this technology its use is becoming increasingly economic.

Non-terrain Applications

For appropriate mathematical functions and especially for certain types of statistics displays, a similar model may be constructed as an aid to understanding a function or as an aid to studying the statistical data.

Notable Examples

The Great Polish Map of Scotland is claimed to be the largest terrain relief model, constructed out of brick and concrete in the grounds of a hotel near Peebles, Scotland. It measures 50 by 40 metres (160 ft × 130 ft).

However, a site in Ningxia province, China was spotted in 2006 using satellite imagery. It measured 900 by 700 metres (3,000 ft × 2,300 ft), had a 3-kilometre (1.9 mi) perimeter and appeared to be a large scale relief model of Aksai Chin, a disputed territory between China and India.

Dymaxion Map

	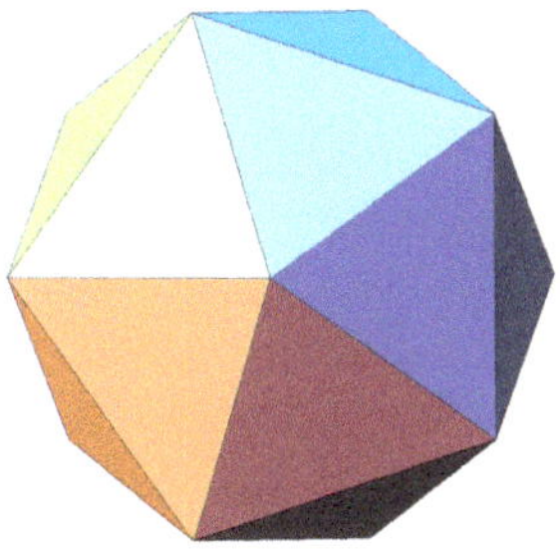	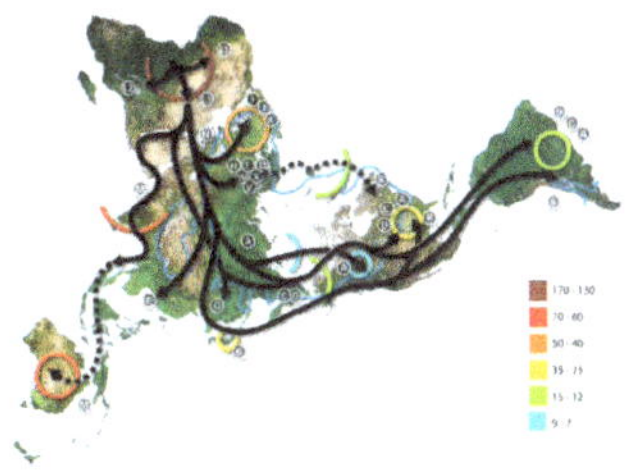
The world is flattened into a Dymaxion map as it unfolds into an icosahedron net with nearly contiguous land masses	An icosahedron: This is the shape onto which the world map is projected before unfolding	Example of use illustrating early human migrations according to mitochondrial population genetics (numbers are millennia before present)

Map of the world in a Fuller projection with Tissot's Indicatrix of deformation	This icosahedral net shows connected oceans surrounding Antarctica	Unfolded Dymaxion map with nearly contiguous land masses

The Dymaxion map or Fuller map is a projection of a world map onto the surface of an icosahedron, which can be unfolded and flattened to two dimensions. The flat map is heavily interrupted in order to preserve shapes and sizes.

The projection was invented by Buckminster Fuller. The March 1, 1943 edition of *Life* magazine included a photographic essay titled "Life Presents R. Buckminster Fuller's Dymaxion World". The article included several examples of its use together with a pull-out section that could be assembled as a "three-dimensional approximation of a globe or laid out as a flat map, with which the world may be fitted together and rearranged to illuminate special aspects of its geography." Fuller applied for a patent in the United States in February 1944, the patent application showing a projection onto a cuboctahedron. The patent was issued in January 1946.

The 1954 version published by Fuller, made with co-cartographer Shoji Sadao, the Airocean World Map, used a modified but mostly regular icosahedron as the base for the projection, which is the version most commonly referred to today. This version depicts the Earth's continents as "one island", or nearly contiguous land masses.

The Dymaxion projection is intended only for representations of the entire globe. It is not a gnomonic projection, whereby global data expands from the center point of a tangent facet outward to the edges. Instead, each triangle edge of the Dymaxion map matches the scale of a partial great circle on a corresponding globe, and other points within each facet shrink toward its middle, rather than enlarging to the peripheries.

The name *Dymaxion* was applied by Fuller to several of his inventions.

Properties

Fuller claimed that his map had several advantages over other projections for world maps.

It has less distortion of relative size of areas, most notably when compared to the Mercator projection; and less distortion of shapes of areas, notably when compared to the Gall–Peters projection. Other compromise projections attempt a similar trade-off.

More unusually, the Dymaxion map does not have any "right way up". Fuller argued that in the universe there is no "up" and "down", or "north" and "south": only "in" and "out". Gravitational forces of the stars and planets created "in", meaning "towards the gravitational center", and "out", meaning "away from the gravitational center". He attributed the north-up-superior/south-down-inferior presentation of most other world maps to cultural bias.

Fuller intended the map to be unfolded in different ways to emphasize different aspects of the world. Peeling the triangular faces of the icosahedron apart in one way results in an icosahedral net that shows an almost contiguous land mass comprising all of Earth's continents – not groups of continents divided by oceans. Peeling the solid apart in a different way presents a view of the world dominated by connected oceans surrounded by land.

Showing the continents as "one island earth" also helped Fuller explain, in his book *Critical Path*, the journeys of early seafaring people, who were in effect using prevailing winds to circumnavigate this world island.

Impact

A 1967 Jasper Johns painting, *Map (Based on Buckminster Fuller's Dymaxion Air-ocean World)*, depicting a Dymaxion map, hangs in the permanent collection of the Museum Ludwig in Cologne.

The World Game, a collaborative simulation game in which players attempt to solve world problems, is played on a 70-by-35-foot Dymaxion map.

In 2013, to commemorate the 70th anniversary of the publication of the Dymaxion map in *Life* magazine, the Buckminster Fuller Institute announced the "Dymax Redux", a competition for graphic designers and visual artists to re-imagine the Dymaxion map. The competition received over 300 entries from 42 countries.

Pictorial Maps

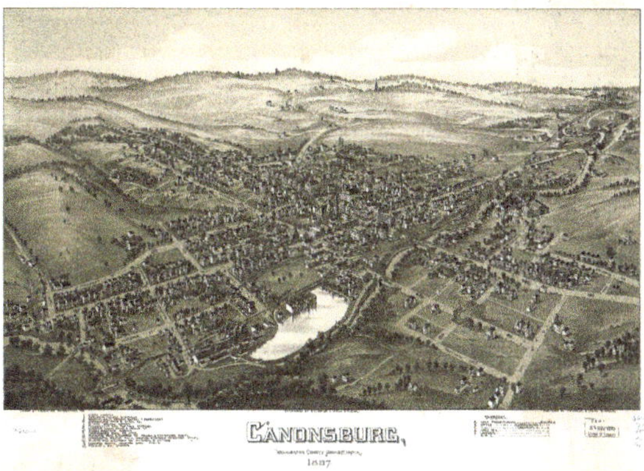

Pictorial map of Canonsburg, Pennsylvania from1897 by Thaddeus Mortimer Fowler & James B. Moyer.

Pictorial maps (also known as illustrated maps, panoramic maps, perspective maps, bird's-eye view maps, and geopictorial maps) depict a given territory with a more artistic rather than technical style. It is a type of map in contrast to road map, atlas, or topographic map. The cartography can be a sophisticated 3-D perspective landscape or a simple map graphic enlivened with illustrations of buildings, people and animals. They can feature all sorts of varied topics like historical events, legendary figures or local agricultural products and cover anything from an entire continent to a college campus. Drawn by specialized artists and illustrators, pictorial maps are a rich, centuries-old tradition and a diverse art form that ranges from cartoon maps on restaurant placemats to treasured art prints in museums.

Pictorial maps usually show an area as if viewed from above at an oblique angle. They are not generally drawn to scale in order to show street patterns, individual buildings, and major landscape features in perspective. While regular maps focus on the accurate rendition of distances, pictorial maps enhance landmarks and often incorporate a complex interplay of different scales into one image in order to give the viewer a more familiar sense of recognition. With an emphasis on objects and style, these maps cover an artistic spectrum from childlike caricature to spectacular landscape graphic with the better ones being attractive, informative and highly accurate. Some require thousands of hours to produce.

The History and Tradition of Pictorial Maps

Pictorial map of Paris by Claes Jansz. Visscher

Will Durant said that maps show us the face of History. This is especially true of pictorial maps because their vocation has always been to present a visual message. Throughout the ages, pictorial maps have been used to show the cuisine of a country, the industries of a city, the attractions of a tourist town, the history of a region or its holy shrines.

The history of pictorial maps overlaps much with the history of cartography in general and ancient artifacts suggest that pictorial mapping has been around since recorded history began.

In Medieval cartography, pictorial icons as well as religious and historical ideas usually overshadowed accurate geographic proportions. A classic example of this is the T and O map which represented the three known continents in the form of a cross with Jerusalem at its center. The more precise art of illustrating detailed bird's-eye-view urban landscapes flourished during the European Renaissance. As emerging trade centers such as Venice began to prosper, local rulers commissioned artists to develop pictorial overviews of their towns to help them organize trade fairs and direct the increasing flow of visiting merchants. When printing came around, pictorial maps evolved into some of the earliest forms of advertising as cities competed amongst themselves to attract larger shares of the known world's commerce.

Later, during the Age of Exploration, maps became progressively more accurate for navigation needs and were often sprinkled with sketches and drawings such as sailing ships showing the direction of trade winds, little trees and mounds to represent forests and mountains and of course, plenty of sea creatures and exotic natives much of them imaginary. As the need for geographical accuracy increased, these illustrations gradually slipped off the map and onto the borders and eventually disappeared altogether in the wake of modern scientific cartography.

The 19th Century

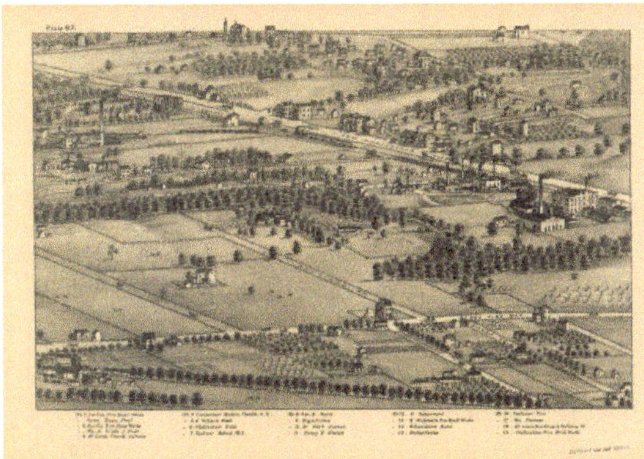

A pictorial map plate of a rural and industrial area in St. Louis during the 19th century

As cartography evolved, the pictorial art form went its own way and regained popularity in the 19th century with the development of the railroads. Between 1825 and 1875, the production and collection of panoramic maps of cities rose to something of a mania. In the U.S. alone, thousands of panoramic maps were produced. The leading panoramic map artists in the U.S.A. were Herman Brosius, Camille N. Drie, Thaddeus Mortimer Fowler, Paul Giraud, Augustus Koch, D. D. Morse, Henry Welge, and A. L. Westyard. Somewhat like the websites of their time, every town had to have one to remain competitive in attracting industry and the immigrant trade. Sometimes artistic exaggeration bordered on the fraudulent as some travelers were drawn by images of idyllic, bustling towns with humming factories only to find a sad little bunch of mud-soaked shacks when they got there. A vast collection of these prints is maintained by the Library of Congress and many of the more beautiful ones continue to be reprinted and sold to this day.

The 20th Century

With the growth of tourism, pictorial mapmaking reappeared as a popular culture art form in the 1920s through the 1950s, often with a whimsical Art Deco style that reflects the period.

Another resurgence occurred in the 1970s and 80s. This was the heyday of companies like Archar and Descartes who produced hundreds of colorful promotional maps of mainly American and Canadian cities. Local businesses were flatteringly drawn on these 'Character maps' with their logos proudly embedded on their buildings. Looking at these maps and who sponsored them over the years, one can clearly see the changing face of industry as the dominant illustrations of manufacturing plants gave way to those of business parks and logos of the service and high tech economy.

Today, like in all other forms of media, the digital revolution has changed the way pictorial maps are researched and executed. But like good writing, a good pictorial map is always a result of intricate labor, esthetic choices and creative editing rather than technology. In the era when Google Earth can give us fly-over access to nearly any spot on the globe, it is amazing to realize that many of the beautiful prints of yore were executed before there were airplanes or even cameras. Whether drawn with quill, pen or pixels, pictorial maps are always scaled in the perspective of the imagination.

Pictorial Map-makers up to Modern Times

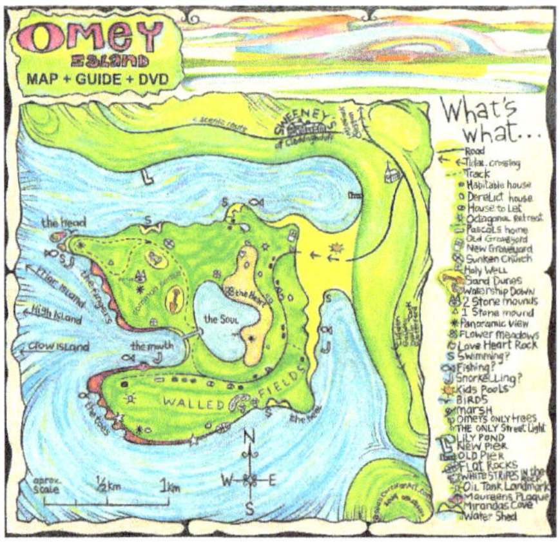

Colourful quirky map of Omey Island created by Irish Artist Sean Corcoran 2009

Ironically, despite all the changes that they record, very little has changed in the business of creating pictorial maps over the centuries. Showing off a given town, attracting visitors and stirring up local pride is what they have always been about. Most of these maps were and continue to be created by a handful of itinerant specialists who keep up the tradition. Many of them traveled from city to city enlisting the support of local merchants, industrialists and civic organizations whose endorsement would of course guarantee a prominent place for their properties on the map.

Edwin Whitefield for instance, one of the more prolific 19th-century American pictorial map artists, would require about 200 subscribers before he put pen to paper. Once he secured the profitability of the venture, Whitefield would be seen all over town furiously sketching every building. Then, choosing an imaginary aerial vantage point, he would integrate all his sketches into a complete and detailed drawing of the city. Then after that, say the chroniclers of the time, Whitefield would once again be seen furiously darting all over town to collect from all his sponsors. Says Jean-Louis Rheault, a contemporary pictorial map illustrator: 'Pictorial maps - with their emphasis on what's important and eye-catching - make it easier to figure out what's where.'.

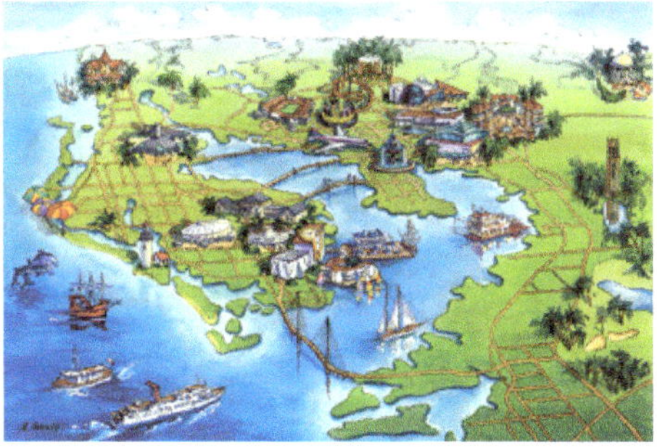

Tampa-Bay Aerial View Map by Maria Rabinky 2008

Fantasy Map

A fantasy map of the Island of Sodor showing the railway system

A fantasy map is type of map design that is a visual representation of an imaginary or fictional geography. While some fantasy maps accompany works of fiction and are considered fictional maps, fantasy maps are created to show imaginary places and are not necessarily included in works of literary fiction. Depending on the completeness and complexity of the map, the depiction of geographical components can range from simple drawings of a small area as in *The Twenty-One Balloons* by William Pène du Bois to an entire fictional world as in *The Lord of the Rings* by Tolkien to even an entire galaxy as in *Star Trek*. Fantasy maps can also include abstract works of art, combine existing cartographic information to present an imaginary location, or combine existing cartographic information to show a different perspective of a location.

In Popular Culture

- Lewis Carroll's *Sylvie and Bruno Concluded* tells of a fantasy map that had "the scale of a mile to the mile." It was later referenced (possibly independent of Carroll) by comedian Steven Wright, who claimed that it was not detailed enough and planned to obtain another in which one foot of space in real life was represented by three feet on the map.
- Jorge Luis Borges wrote a short story about a map made to 1:1 scale. It is an homage to Lewis Carroll's work mentioned above. The story, "On Exactitude in Science", is collected in *A Universal History of Infamy*.

- J. R. R. Tolkien's Middle-earth is a renowned example of a fantasy map.
- Nikolas Schiller's kaleidoscopic aerial photography of urban areas

City Map

A city map is a large-scale thematic map of a city (or part of a city) created to enable the fastest possible orientation in an urban space. The graphic representation of objects on a city map is therefore usually greatly simplified, and reduced to generally understood symbology.

City map of Stavanger (Norway).

Depending upon its target group or market, a city map will include not only the city's transport network, but also other important information, such as city sights or public institutions.

Content and Design

The scale of a city map is usually between 1:10,000 and 1:25,000. Densely settled downtown areas will sometimes be partly drawn in a larger scale, on a separate detail map.

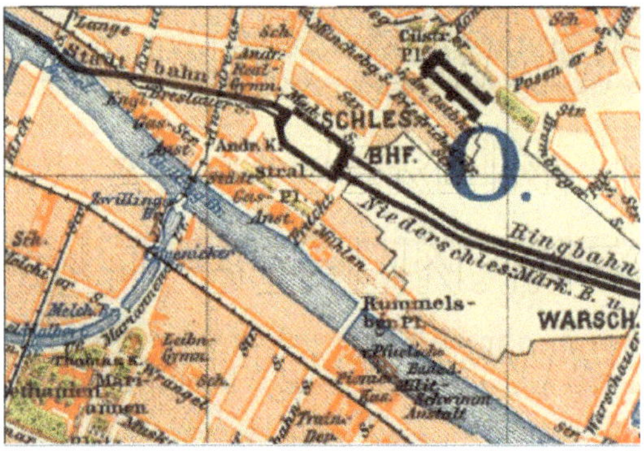

Detail from a city map of Berlin from 1895.

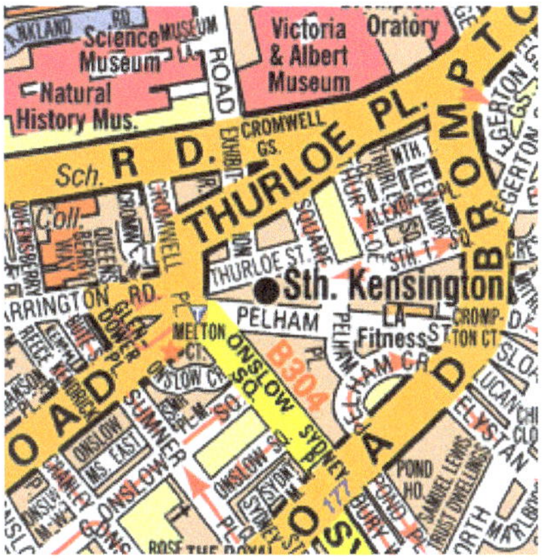

Detail of Geographers' A-Z Street Atlas of London.

In addition to linear true to scale maps, there are also maps with variable scale, for example where the scale gradually increases towards the city centre (a fish-eye projection).

Central to the information provided by a city map is the street network, including its street names (often supplemented by at least a selection of individual house numbers), along with buildings, parks and waterways. Streets and points of interest are usually also listed in a legend or register, locating objects on a map grid on the map. Important places such as administrative buildings, cultural institutions, attractions, etc. may be highlighted with the assistance of pictograms. The map may also be complemented by representations of public transport facilities.

History

Clay tablet with map of the Babylonian city of Nippur (ca. 1400 BC)

Ancient Near East

As early as the time of the Ancient Near East, clay tablets were being produced with scaled, graphical representations of cities. Excavations of the Sumerian city of Nippur brought to light a fragment of an approximately 3,500-year-old city map, which is sometimes referred to as the oldest known city map. The clay tablet depicts the temple of Enlil, a city park, the city wall including its gates, along with a canal and the river Euphrates. The individual objects on this map were already labelled, in a Sumerian cuneiform.

Late Middle Ages

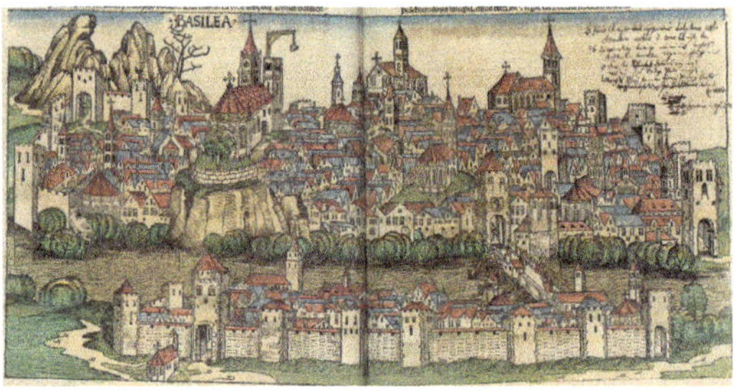

View of Basel, Switzerland, ca 1490, from the Nuremberg Chronicle.

In manuscripts and early printed books of the Late Middle Ages, cities are often shown in profile, or viewed from an elevated standpoint. Nautical charts of that time sometimes depict partly stylized cityscapes drawn in pictogram form - for example in Cristoforo Buondelmonti's *Liber insularum archipelagi* (Book of Islands), from the year 1422.

The Nuremberg Chronicle, which first appeared in 1493, is one of the most important collections of city views of the late Middle Ages, with over 100 such illustrations. Nevertheless, panoramas like this one, or the one in Bernhard von Breydenbach's *Travelogue* from 1483, had more narrative or representative functions.

Illustrated in these works are the local conditions and relevant characteristics - such as ports, magnificent buildings, walls, etc. - as a background for highlighting historical descriptions or economic benefits of the city. On the other hand, less emphasis was placed on accuracy: in the Nuremberg Chronicle, merely one quarter of the city views represented the actual appearance of the subject city, and some individual images were even used simultaneously to represent several different cities.

Renaissance

Antwerp (engraving), ca 1572, by Georg Braun and Frans Hogenberg.

In the 16th century, the artists and scholars of the Renaissance had extensive knowledge of mathematical perspectives and projections. This knowledge also affected the work of cartographers and the production of cityscapes (especially in Italy). A key innovation was that the city was no longer portrayed simply from an imaginary or real perspective, but drawn initially as a two-dimensional map, and then, using a process of accurate perspective drawing, transformed into a three-dimensional image. An early example of a geometrically exact and highly detailed work of this kind is the city map of Venice created by Jacopo de' Barbari in around 1500.

Whereas the illustrations of the late Middle Ages are usually still simple small-format woodcuts, an increasingly common process from 1500 onwards was the creation of prints from huge woodcuts and woodblocks. Jacopo de' Barbari's map of Venice was already as large as 139 centimetres (55 in) x 282 centimetres (111 in), and consisted of six individual panels. From the middle of the 16th century, the copperplate process, originating in Antwerp, began to compete with the woodcut, and allowed far more refined and detailed illustrations.

19th Century Pocket Atlases

One of the first city pocket atlases, and the first pocket atlas of London, was "Collins' Illustrated Atlas of London" published in 1854 and drawn and engraved by Richard Jarman.

Topographic Map

In modern mapping, a topographic map is a type of map characterized by large-scale detail and quantitative representation of relief, usually using contour lines, but historically using a variety of methods. Traditional definitions require a topographic map to show both natural and man-made features. A topographic map is typically published as a map series, made up of two or more map sheets that combine to form the whole map. A contour line is a line connecting places of equal elevation.

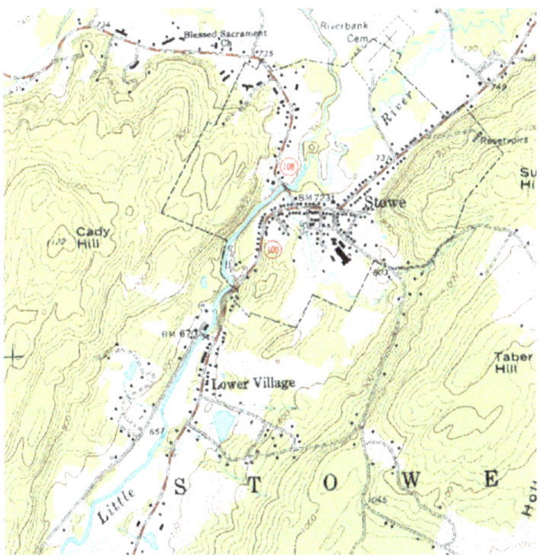

A topographic map with contour lines

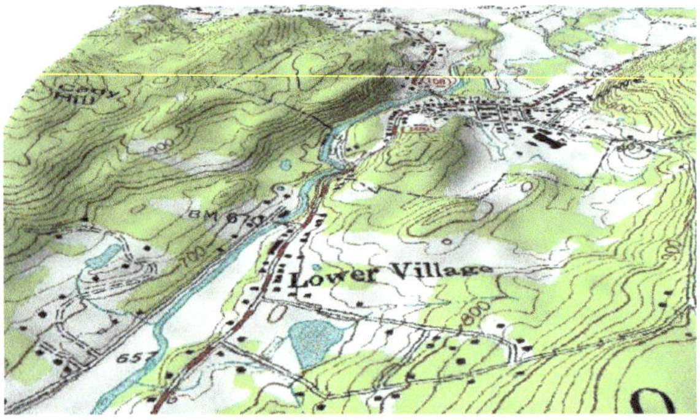

Part of the same map in a perspective shaded relief view illustrating how the contour lines follow the terrain

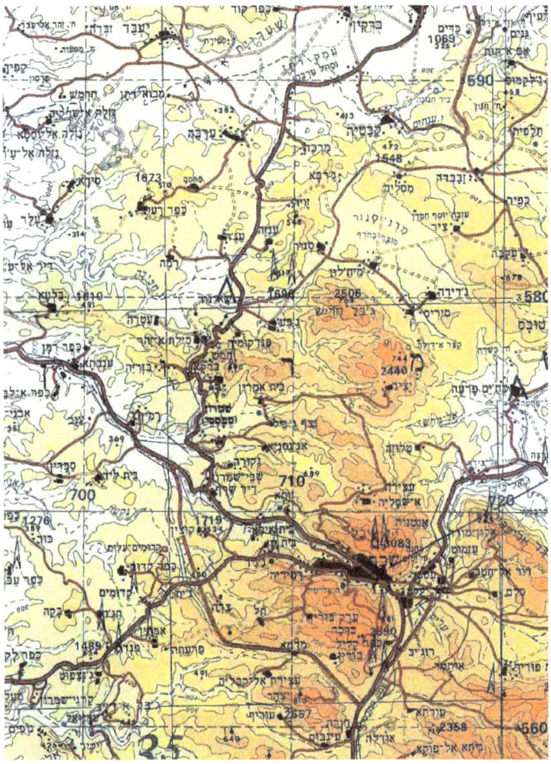

Section of topographical map of Nablus area (West Bank) with contour lines at 100-meter intervals. Heights are colour-coded

Natural Resources Canada provides this description of topographic maps:

These maps depict in detail ground relief (landforms and terrain), drainage (lakes and rivers), forest cover, administrative areas, populated areas, transportation routes and facilities (including roads and railways), and other man-made features.

Other authors define topographic maps by contrasting them with another type of map; they are distinguished from smaller-scale "chorographic maps" that cover large regions, "planimetric maps" that do not show elevations, and "thematic maps" that focus on specific topics.

However, in the vernacular and day to day world, the representation of relief (contours) is popularly held to define the genre, such that even small-scale maps showing relief are commonly (and erroneously, in the technical sense) called "topographic".

The study or discipline of topography is a much broader field of study, which takes into account all natural and man-made features of terrain.

History

Topographic maps are based on topographical surveys. Performed at large scales, these surveys are called topographical in the old sense of topography, showing a variety of elevations and landforms. This is in contrast to older cadastral surveys, which primarily show property and governmental boundaries. The first multi-sheet topographic map series of an entire country, the *Carte géométrique de la France*, was completed in 1789. The Great Trigonometric Survey of India, started by the East India Company in 1802, then taken over by the British Raj after 1857 was notable as a successful effort on a larger scale and for accurately determining heights of Himalayan peaks from viewpoints over one hundred miles distant.

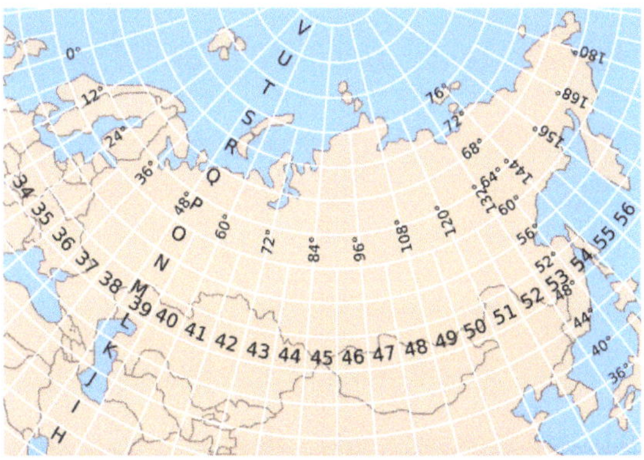

Global indexing system first developed for *International Map of the World*

Topographic surveys were prepared by the military to assist in planning for battle and for defensive emplacements (thus the name and history of the United Kingdom's Ordnance Survey). As such, elevation information was of vital importance.

As they evolved, topographic map series became a national resource in modern nations in planning infrastructure and resource exploitation. In the United States, the national map-making function which had been shared by both the Army Corps of Engineers and the Department of the Interior migrated to the newly created United States Geological Survey in 1879, where it has remained since.

1913 saw the beginning of the International Map of the World initiative, which set out to map all of Earth's significant land areas at a scale of 1:1 million, on about one thousand sheets, each covering four degrees latitude by six or more degrees longitude. Excluding borders, each sheet was 44 cm high and (depending on latitude) up to 66 cm wide. Although the project eventually foundered, it left an indexing system that remains in use.

By the 1980s, centralized printing of standardized topographic maps began to be superseded by databases of coordinates that could be used on computers by moderately skilled end users to view or print maps with arbitrary contents, coverage and scale. For example, the Federal government of the United States' *TIGER* initiative compiled interlinked databases of federal, state and local political borders and census enumeration areas, and of roadways, railroads, and water features with support for locating street addresses within street segments. TIGER was developed in the 1980s and used in the 1990 and subsequent decennial censuses. Digital elevation models (*DEM*) were also compiled, initially from topographic maps and stereographic interpretation of aerial photographs and then from satellite photography and radar data. Since all these were government projects funded with taxes and not classified for national security reasons, the datasets were in the public domain and freely usable without fees or licensing.

TIGER and DEM datasets greatly facilitated Geographic information systems and made the Global Positioning System much more useful by providing context around locations given by the technology as coordinates. Initial applications were mostly professionalized forms such as innovative surveying instruments and agency-level GIS systems tended by experts. By the mid-1990s, increasingly user-friendly resources such as online mapping in two and three dimensions, integration of GPS with mobile phones and automotive navigation systems appeared. As of 2011, the future of standardized, centrally printed topographical maps is left somewhat in doubt.

Uses

Curvimeter used to measure the length of a curve

Topographic maps have multiple uses in the present day: any type of geographic planning or large-scale architecture; earth sciences and many other geographic disciplines; mining and other earth-based endeavours; civil engineering and recreational uses such as hiking and orienteering.

Conventions

The various features shown on the map are represented by conventional signs or symbols. For example, colors can be used to indicate a classification of roads. These signs are usually explained in the margin of the map, or on a separately published characteristic sheet.

Topographic maps are also commonly called *contour maps* or *topo maps*. In the United States, where the primary national series is organized by a strict 7.5-minute grid, they are often called *topo quads* or quadrangles.

Topographic maps conventionally show topography, or land contours, by means of contour lines. Contour lines are curves that connect contiguous points of the same altitude (isohypse). In other words, every point on the marked line of 100 m elevation is 100 m above mean sea level.

These maps usually show not only the contours, but also any significant streams or other bodies of water, forest cover, built-up areas or individual buildings (depending on scale), and other features and points of interest.

Today, topographic maps are prepared using photogrammetric interpretation of aerial photography, lidar and other Remote sensing techniques. Older topographic maps were prepared using traditional surveying instruments.

Publishers of National Topographic Map Series

Although virtually the entire terrestrial surface of Earth has been mapped at scale 1:1,000,000, medium and large-scale mapping has been accomplished intensively in some countries and much less in others. Nevertheless, national mapping programs listed below are only a partial selection. Several commercial vendors supply international topographic map series.

Australia

The NMIG (National Mapping Information Group) of Geoscience Australia is the Australian Government's national mapping agency. It provides topographic maps and data to meet the needs of the sustainable development of the nation. The Office of Spatial Data Management provides an online free map service MapConnect. These topographic maps of scales 1:250,000 and 1:100,000 are available in printed form from the Sales Centre. 1:50,000 and 1:25,000 maps are produced in conjunction with the Department of Defence.

Austria

Austrian Maps (German: Österreichische Karte (ÖK)) is the government agency producing maps of Austria, which are distributed by *Bundesamt für Eich- und Vermessungswesen* (BEV) in Vienna. The maps are published at scales 1:25,000 1:50,000 1:200,000 and 1:500,000. Maps can also be viewed online.

Canada

The Centre for Topographic Information produces topographic maps of Canada at scales of 1:50,000 and 1:250,000. They are known as the National Topographic System (NTS). A government proposal to discontinue publishing of all hardcopy or paper topographic maps in favor of digital-only mapping data was shelved in 2005 after intense public opposition.

China

The *State Bureau of Surveying and Cartography* compiles topographic maps at 1:25,000 and 1:50,000 scales. It is reported that these maps are accurate and attractively printed in seven colors, and that successive editions show progressive improvement in accuracy.

These large-scale maps are the basis for maps at smaller scales. Maps at scales 1:4,000,000 or smaller are exported by *Cartographic Publishing House, Beijing* while larger-scale maps are restricted as state secrets, and prohibited from publishing by legislation, all except Hong Kong and Macau. China's topographic maps follow the international system of subdivision with 1:100,000 maps spanning 30 minutes longitude by 20 minutes latitude.

Colombia

The Geographic Institute Agustín Codazzi is the government entity responsible for producing and distributing topographic maps of Colombia in 1:500,000 and 1:100,000 scales. These and several other Geographic information services can be accessed using the Instituto Geográfico Agustin Codazzi website in Spanish.

Denmark

The National Survey and Cadastre of Denmark is responsible for producing topographic and nautical geodata of Denmark, Greenland and the Faroe Islands.

Finland

The National Land Survey of Finland produces the Topographic Database (accuracy 1:5000-1:10 000) and publishes topographic maps of Finland at 1:25,000 and 1:50,000. In addition topographics maps can be viewed by using a free map service MapSite.

France

The Institut Géographique National (IGN) produces topographic maps of France at 1:25,000 and 1:50,000. In addition, topographic maps are freely accessible online, through the Géoportail website.

Germany

In principle, each federal state *(Bundesland)* is in charge of producing the official topographic maps. In fact, the maps between 1:5,000 and 1:100,000 are produced and published by the land surveying offices of each federal state, the maps between 1:200,000 and 1:1,000,000 by a federal office – the Bundesamt für Kartographie und Geodäsie (BKG) in Frankfurt am Main.

Greece

Topographic maps for general use are available at 1:50,000 and 1:100,000 from the *Hellenic Military Geographical Service (HMGS).* They use a national projection system called EGSA'87, which is a Transverse Mercatorial Projection mapping Gre ece in one zone. A few areas are also available at 1:25,000. Some private firms sell topographic maps of national parks based on HMGS topography.

Hong Kong

The Department of Lands is the government agency responsible for surveying and publishing topographic maps of Hong Kong. Commonly used maps such as the HM20C series (1:20,000) are reviewed and updated every year or two. Very large scale (1:600 in Urban area and the 1:1,000 HM1C series for all of HK) topographic maps are also available to public for various uses.

India

The Survey of India is responsible for all topographic control, surveys and mapping of India.

Japan

The Geographical Survey Institute of Japan is responsible for base mapping of Japan. Standard map scales are 1:25,000, 1:50,000, 1:200,000 and 1:500,000.

Nepal

From 1992 to 2002 a definitive series of large scale topographic maps were surveyed and published through a joint project by Government of Nepal Survey Department and Finland's Ministry for Foreign Affairs contracting through the *FinnMap* consulting firm. Japan International Cooperation Agency substituted for FinnMap in Lumbini Zone.

Topographic sheets at 1:25,000 scale covering 7.5 minutes latitude and longitude map the densely populated*Terai* and *Middle Mountain* regions. Less populated high mountain regions are on 15-minute sheets at 1:50,000. JPG scans can be downloaded.

Netherlands

The Land Registry Kadaster (formerly *Topografische Dienst*) collects, processes and provides topographical information of the entire Dutch territory. The history of the Land Registry goes back to the year 1815, that year was commissioned to create a large map, known as *Map of Krayenhoff*. Around 1836 they began printing the topographic map on a scale of 1: 50,000, followed in 1865 by the topographic map on a scale of 1: 25,000. In 1951 began the start of production of the topographic map on a scale of 1: 10,000. From various reorganizations arose in 1932 the *Topografische Dienst* as national mapping agency of the Netherlands, since January 2004 housed within the Land Registry *Kadaster*.

New Zealand

Land Information New Zealand is the government agency responsible for providing up-to-date topographic mapping. LINZ topographic maps cover all of New Zealand, offshore islands, some Pacific Islands and the Ross Sea Region. The standard issue *NZTopo* map series was published September 2009 at 1:50,000 (NZTopo50), and 1:250,000 (NZTopo250). Vector data from the New Zealand Topographic Database (NZTopo) is also available.

Pakistan

The responsibility for topographic mapping and aerial photography lies with the Surveyor General of Pakistan [SGP]. Established in 1947, the Survey of Pakistan (SOP) is based in Rawalpindi with a number of regional offices distributed at urban centers throughout Pakistan. SGP is a civil organization which, for security reasons, is headed by a Surveyor General and works under the strict control of Army General Headquarters (GHQ). Colonel C.A.K. Innes-Wilson, a Royal Engineers officer who joined the Survey of India which mapped the subcontinent, was the first Surveyor General of Pakistan.

All departments which require topographic maps make their request to SGP and many are permanently registered with it for mapping and aerial photographs procurement. The SOP performs these functions under the auspices of the Ministry of Defence (MOD). Organisationally, the SOP is overseen by the Surveyor General (SG) who is a direct military appointee and a senior uniformed officer. The SG reports directly to the Secretary of Defence. Under the SG are two Deputy SG's (I and II) who manage the operational departments of the agency and a Senior Technical Advisor. These departments are divided into Regional Directorates for Topographic Mapping including the Northern region centred in Peshawar, Eastern region (Lahore), Western region (Quetta) and finally, the Southern region in Karachi. Responsibility for fields surveys and the maintenance/update of topographic maps are sub-divided according to these geographic areas.

Portugal

The Army's Geographical Institute - Instituto Geográfico do Exército - produces 1.25,000, 1:500,000 maps for public sale, as well as lots of geographical services.

Romania

Until recently, the two major government mapping authorities in Romania have been the Military Topographic Department (Directia Topografica Militara (DTM)), and the Institute for Geodesy, Photogrammetry, Cartography and Land Management (Institutul de Geodezie, Fotogrammetrie, Cartografie, si Organizarea Teritoriului (IGFCOT)). This situation has recently changed, following a decision in 1996 by the Romanian Government to establish a combined civilian National Office of Cadastre, Geodesy and Cartography (Oficiul National de Cadastru, Geodezie si Cartografie (ONCGC). Maps continued to be published under the imprint of the previous organizations into the late 1990s. From 1958, a number of town maps at scales of 1:5,000 or 1:10,000 were also made, initially on the Gauss-Krüger projection, but after 1970 on a stereographic projection. More than 100 such sheets have been produced. There is also a street map of Bucharest in four sheets at 1:15,000 derived from larger scale surveys, which is revised annually.

The 1:50,000 series in 737 sheets is now regarded as the base map. It was revised in the period 1965-72 using aerial photographs, and is currently being updated again with the intention of establishing a revision cycle of five to six years. The 1:25,000 will be retained, but revision only at 15-20-year intervals, except for sheets covering areas of rapid change.

Russia

Detailed, accurate topographic maps have long been a military priority. They are currently produced by the *Military-topographic service of armed forces of the Russian Federation* (Russian: Военно-топографическая служба Вооружённых сил Российской Федерации or ВТС ВС). Military topographic mapping departments held other titles in the Russian Empire since 1793 and in the Soviet Union where these maps also came to be used for internal control and economic development.

When Germany invaded in 1941, detailed maps from the USSR's western borders to the Volga River became an urgent task, accomplished in less than one year. After the war years the entire Soviet Union was mapped at scales down to 1:25,000—even 1:10,000 for the agriculturally productive fraction. The rest of the world except Antarctica is believed to have been mapped at scales down to 1:200,000, with regions of special interest down to 1:50,000 and many urban areas to 1:10,000. In all there may have been over one million map sheets of high quality and detail. Soviet maps were also notable for their consistent global indexing system. These advantages held for Soviet military maps of other countries, although there were some errors due to faulty intelligence.

Soviet maps for domestic civilian purposes were often of lower quality. From 1919 to 1967 they were produced by *Head geodesic administration* (Russian: Высшее геодезическое управление or ВГУ), then by *Chief administration of geodesy and cartography* (Russian: Главное управление геодезии и картографии or ГУГК). Now (June 2011) civilian maps are produced by the *Federal agency for geodesy and cartography* (Russian: Федеральное агентство геодезии и картографии or Роскартография).

Soviet military maps were state secrets. After the 1991 breakup of the Soviet Union, many maps leaked into the public domain and are available for download. Map scales 1:100.000 - 1:500.000 can be viewed online.

Spain

The Instituto Geográfico Nacional (IGN) is responsible for the official topographic maps. It does use six scales that cover all the Spanish territory: 1:25,000, 1:50,000, 1:200,000, 1:500,000, 1:1,000,000 and 1:2,000,000. The most common scale is the first one, which utilizes the UTM system.

South Africa

The Chief Directorate: National Geo-spatial Information (CD:NGI) produces three topographic map series, each covering the whole country, at scales 1:50 000, 1:250 000, and 1:500 000.

Switzerland

Swisstopo (the Federal Office of Topography) produces topographic maps of Switzerland at seven different scales.

Taiwan

Topographic maps for Taiwan had long been kept as confidential information due to security concerns. It has only been recently made available to public from the National Land Surveying and Mapping Center, the government agency responsible for surveying and publishing various maps. Topographic maps of up to 1:25,000 is now available in digital and printed format.

United Kingdom

The Ordnance Survey (OS) produces topographic map series covering the United Kingdom at 1:25,000 and 1:50,000 scales. The 1:25,000 scale is known as the "Explorer" series, and include an "OL" (Outdoor Leisure) sub-series for areas of special interest to hikers and walkers. It replaced the "Pathfinder" series, which was less colourful and covered a smaller area on each map. The 1:50,000 scale is known as the "Landranger" and carries a distinctive pink cover. More detailed mapping as fine as 1:10,000 covers some parts of the country. The 1:25,000 and 1:50,000 scales are easily coordinated with standard romer scales on currently available compasses and plotting tools. The Ordnance Survey maintains a mapping database from which they can print specialist maps at virtually any scale.

The Ordnance Survey National Grid divides the U.K. into cells 500 km, 100 km, 10 km and 1 km square on a Transverse Mercator grid aligned true North-South along the 2°W meridian. OS map products are based on this grid.

United States

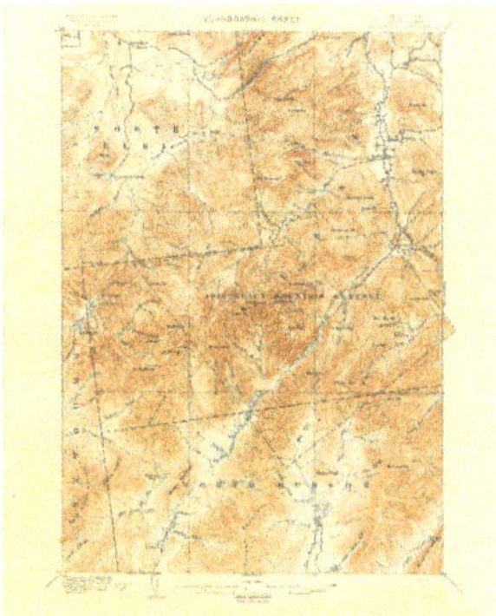

The Mount Marcy area of New York State in 1892 in a 15-minute quadrangle at 1:62,500.

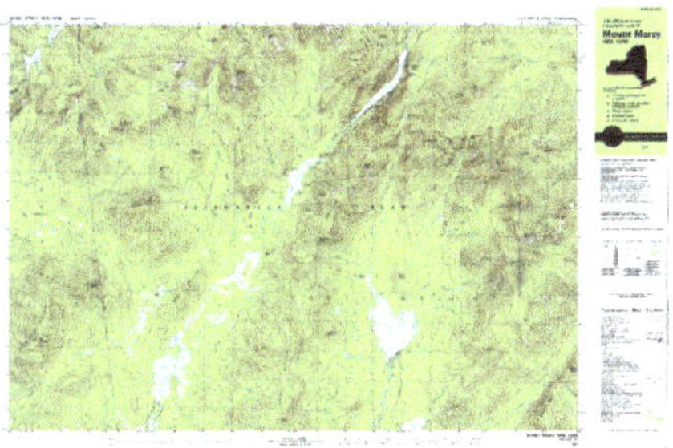

The same area about a century later (1979) in a 7.5- by 15-minute metric map at 1:25,000.

The United States Geological Survey (USGS), a civilian federal agency, produces several national series of topographic maps which vary in scale and extent, with some wide gaps in coverage, notably the complete absence of 1:50,000 scale topographic maps or their equivalent. The largest (both in terms of scale and quantity) and best-known topographic series is the 7.5-minute or 1:24,000 quadrangle. This scale is unique to the United States, where nearly every other developed nation has introduced a metric 1:25,000 or 1:50,000 large scale topo map. The USGS also publishes 1:100,000 maps covering 30 minutes latitude by one degree longitude, 1:250,000 covering one by two degrees, and state maps at 1:500,000 with California, Michigan and Montana needing two sheets while Texas has four. Alaska is mapped on a single sheet, at scales ranging from 1:1,584,000 to 1:12,000,000.

Recent USGS digital *US Topo* 1:24,000 topo maps based on the *National Map* omit[*why?*] several important geographic details that were featured in the original USGS topographic map series (1945-1992). Examples of omitted details and features include power transmission lines, telephone lines, railroads, recreational trails, pipelines, survey marks, and buildings. For many of these feature classes, the USGS is working with other agencies to develop data or adapt existing data on missing details that will be included in The National Map and to US Topo. In other areas USGS digital map revisions may omit geographic features such as ruins, mine locations, springs, wells, and even trails in an effort to protect natural resources and the public at large, or because such features are not present in any public domain database.

Geologic Map

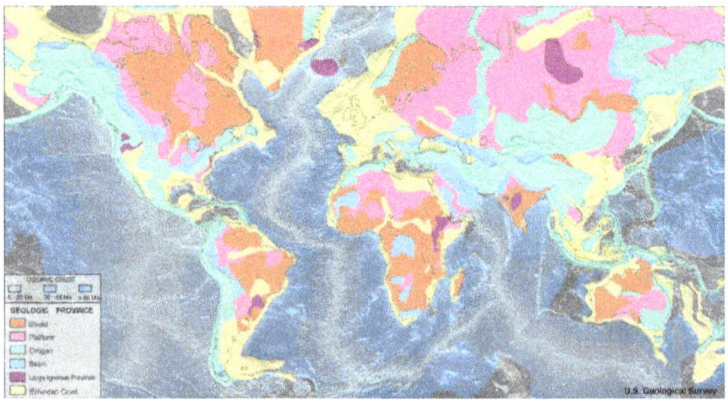

Mapped global geologic provinces

A geologic map or geological map is a special-purpose map made to show geological features. Rock units or geologic strata are shown by color or symbols to indicate where they are exposed at the surface. Bedding planes and structural features such as faults, folds, foliations, and lineations are shown with strike and dip or trend and plunge symbols which give these features' three-dimensional orientations.

William Smith's geologic map

Stratigraphic contour lines may be used to illustrate the surface of a selected stratum illustrating the subsurface topographic trends of the strata. Isopach maps detail the variations in thickness of stratigraphic units. It is not always possible to properly show this when the strata are extremely fractured, mixed, in some discontinuities, or where they are otherwise disturbed.

Lithologies

Rock units are typically represented by colors. Instead of (or in addition to) colors, certain symbols can be used. Different geologic mapping agencies and authorities have different standards for the colors and symbols to be used for rocks of differing types and ages.

Orientations

Geologists take two major types of orientation measurements (using a hand compass like a Brunton compass): orientations of planes and orientations of lines. Orientations of planes are often measured as a "strike" and "dip", while orientations of lines are often measured as a "trend" and "plunge".

Strike and dip symbols consist of a long "strike" line, which is perpendicular to the direction of greatest slope along the surface of the bed, and a shorter "dip" line on side of the strike line where the bed is going downwards. The angle that the bed makes with the horizontal, along the dip direction, is written next to the dip line. In the azimuthal system, strike and dip are often given as "strike/dip" (for example: 270/15, for a strike of west and a dip of 15 degrees below the horizontal).

Trend and plunge are used for linear features, and their symbol is a single arrow on the map. The arrow is oriented in the downgoing direction of the linear feature (the "trend") and at the end of the arrow, the number of degrees that the feature lies below the horizontal (the "plunge") is noted. Trend and plunge are often notated as PLUNGE → TREND (for example: 34 → 86 indicates a feature that is angled at 34 degrees below the horizontal at an angle that is just East of true South).

History

The oldest preserved geologic map is the Turin papyrus (1150 BCE), which shows the location of building stone and gold deposits in Egypt.

The earliest geologic map of the modern era is the 1771 "Map of Part of Auvergne, or figures of, The Current of Lava in which Prisms, Balls, Etc. are Made from Basalt. To be used with Mr. Demarest's theories of this hard basalt. Engraved by Messr. Pasumot and Daily, Geological Engineers of the King." This map is based on Nicolas Desmarest's 1768 detailed study of the geology and eruptive history of the Auvergne volcanoes and a comparison with the columns of the Giant's Causeway of Ireland. He identified both landmarks as features of extinct volcanoes. The 1798 report was incorporated in the 1771 (French) Royal Academy of Science compendium.

The first geological map of the U.S. was produced in 1809 by William Maclure. In 1807, Maclure undertook the self-imposed task of making a geological survey of the United States. He traversed and mapped nearly every state in the Union. During the rigorous two-year period of his survey, he crossed and recrossed the Allegheny Mountains some 50 times. Maclure's map shows the distribution of five classes of rock in what are now only the eastern states of the present-day US.

The first geologic map of Great Britain was created by William Smith in 1815.

Maps and Mapping Around the Globe

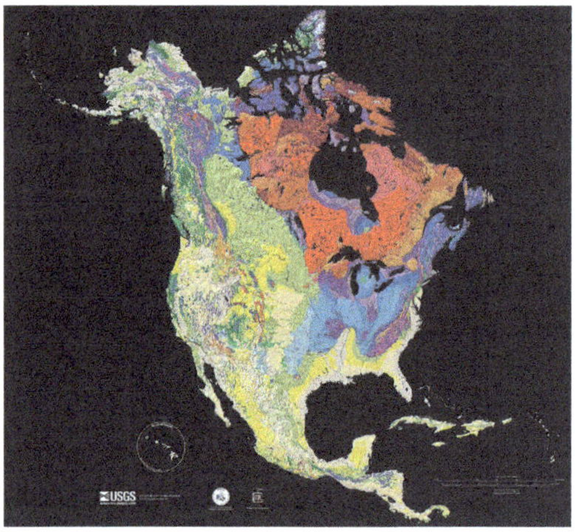

Geologic map of North America superimposed on a shaded relief map

United States

In the United States, geologic maps are usually superimposed over a topographic map (and at times over other base maps) with the addition of a color mask with letter symbols to represent the kind of geologic unit. The color mask denotes the exposure of the immediate bedrock, even if obscured by soil or other cover. Each area of color denotes a geologic unit or particular rock formation (as more information is gathered new geologic units may be defined). However, in areas where the bedrock is overlain by a significantly thick unconsolidated burden of till, terrace sediments, loess deposits, or other important feature, these are shown instead. Stratigraphic contour lines, fault lines, strike and dip symbols, are represented with various symbols as indicated by the map key. Whereas topographic maps are produced by the United States Geological Survey in conjunction with the states, geologic maps are usually produced by the individual states. There are almost no geologic map resources for some states, while a few states, such as Kentucky and Georgia, are extensively mapped geologically. Technically A map that uses colors.

United Kingdom

In the United Kingdom the term *geological map* is used. The UK and Isle of Man have been extensively mapped by the British Geological Survey (BGS) since 1835; a separate Geological Survey of Northern Ireland (drawing on BGS staff) has operated since 1947.

Two 1:625,000 scale maps cover the basic geology for the UK. More detailed sheets are available at scales of 1:250,000, 1:50,000 and 1:10,000. The 1:625,000 and 1:250,000 scales show both onshore and offshore geology (the 1:250,000 series covers the entire UK continental shelf), whilst other scales generally cover exposures on land only.

Sheets of all scales (though not for all areas) fall into two categories:

Superficial deposit maps (previously known as *solid and drift* maps) show both bedrock *and* the deposits on top of it.

Bedrock maps (previously known as *solid* maps) show the underlying rock, without superficial deposits.

The maps are superimposed over a topographic map base produced by Ordnance Survey (OS), and use symbols to represent fault lines, strike and dip or geological units, boreholes etc. Colors are used to represent different geological units. Explanatory booklets (memoirs) are produced for many sheets at the 1:50,000 scale.

Small scale thematic maps (1:1,000,000 to 1:100,000) are also produced covering geochemistry, gravity anomaly, magnetic anomaly, groundwater, etc.

Although BGS maps show the British national grid reference system and employ an OS base map, sheet boundaries are not based on the grid. The 1:50,000 sheets originate from earlier 'one inch to the mile' (1:63,360) coverage utilising the pre-grid Ordnance Survey One Inch Third Edition as the base map. Current sheets are a mixture of modern field mapping at 1:10,000 redrawn at the 1:50,000 scale and older 1:63,360 maps reproduced on a modern base map at 1:50,000. In both cases the original OS Third Edition sheet margins and numbers are retained. The 1:250,000 sheets are defined using lines of latitude and longitude, each extending 1° north-south and 2° east-west.

Singapore

The first geological map of Singapore was produced in 1974, produced by the then Public Work Department. The publication includes a locality map, 8 map sheets detailing the topography and geological units, and a sheet containing cross sections of the island.

Since 1974, for 30 years, there were many findings reported in various technical conferences on new found geology islandwide, but no new publication was produced. In 2006, Defence Science & Technology Agency, with their developments in underground space promptly started a re-publication of the Geology of Singapore, second edition.

The new edition that was published in 2009, contains a 1:75,000 geology map of the island, 6 maps (1:25,000) containing topography, street directory and geology, a sheet of cross section and a locality map.

The difference found between the 1976 Geology of Singapore report include numerous formations found in literature between 1976 and 2009. These include the Fort Canning Boulder Beds and stretches of limestone.

Cartographic Labeling

Cartographic labeling is a form of typography and strongly deals with form, style, weight and size of type on a map. Essentially, labeling denotes the correct way to label features (points, arcs, or polygons).

Form

In type, form describes anything from lengths between letters to the case and color of the font. Form works well for both nominal (qualitative) and ordered (quantitative) data.

Italics

Italics describe the sloping of letters setting it apart from non-italicized words (or vice versa). Using italics on a map also slightly decreases the size of the font as it shapely squeezes it around features. When introduced, the idea was to condense the text by italicizing it, thus creating more text on the pages. The slope in the font was created to mimic the flow of cursive handwriting and thus, the angles of italic letters range anywhere from 11 to 30 degree and consequently, serifs are absent.

As a general rule on maps, the smaller the point size of a font, the more condensed and difficult it becomes to read. In an example of labeling a globe, ocean features are generally italicized to give an obvious discernment. In cartographic conventions, natural features are adequate in italics such as the aforementioned hydrographic features.

Case

Case is another way of emphasizing—whether it be uppercase, lowercase or a combination of the two (or even different size points within the same case). In general, uppercase fonts denote a higher emphasis, but according to Bringhurst (1996), an uppercase initial of a word has the seniority; but the lowercase letters have the control. In other words, the strong boldness of a larger letter draws the audience into its viewpoint. The lowercase letters contain the information needed to convey further. When viewing text on maps, it is still crucial to gain the audience's attention as a way of informing them of something other than the map(s). As for design, uppercase is much harder to read than mixed-use. In the globe example, mountain ranges should be in uppercase. When showing a larger scale, such as a region of the United States, it is useful to classify different case sizes. States should be in uppercase, with counties in small uppercase, and cities in lowercase.

Color

Color (value and hue) alterations also allow for a further emphasis on certain features. By changing the color of the font to correspond to the feature it is representing, the two become joined. If the cartographer were to label a river, the extra emphasis would be inherent if the font chosen was blue, to correspond with the blue feature (arc). On the contrary though, this is not always necessarily the case. If the cartographer chose a color of font for an ocean feature (polygon), blue would not be the obvious choice because it would appear to be washed out and thus, no emphasis. In this case, it is useful to label the feature with a more rich, bolder color (such as black font on blue polygon).

Spacing

The spacing of the letters on features also gives a more appealing map-—visually speaking. By enlarging the increments between each letter of a word, the word in turn, becomes more pronounced. In the case for a long arc feature (river), to add more emphasis on the label, the letters would need to be extended or stretched. On the other hand, in some cases, the letters would have to be condensed (shortened increment gaps) to give a more proportional label for a feature.

Style

Serifs

The type style affects to overall look of the map and is adequately used to symbolize nominal (qualitative) data within the map. In general, style amounts to the use of serifs versus sans serifs. A serif is, by definition, a cross-line at the end of a stroke along a letter. On a map, the text that is chosen should be consistent. Generally, serif fonts are utilized to give a more regimented block body of text—similar to those used in traditional printing. Serifs are more widely used for historical information or a historical map.

Sans Serifs

The serif counterpart is sans serifs (meaning without serifs). Sans serif fonts are the more modern of the two fonts. But choosing one over the other requires that the audience will be able to read the text without strain. Generally, sans serifs are not for large bodies of text in print but instead, are ideal for the internet. On the same facet, sans serifs are optimal for a more-clean appearance in such places like a header, title, or legend. In map design, it's useful to also use sans serifs for natural features.

Weight

The type weight provides a substantial amount of emphasis of the cartographer's choosing. Weight is important because it involves the difference between bold and regular contrast. The degree of power that is increased with weight, must be proportional to the size of the letter. If not, a letter can be too intense and thus more difficult to read. Similarly, the spacing between the letters must be extended to provide adequate to read smoothly. Bold text creates direct attention to the eyes of the audience to pronounce certain information from cartographer.

Size

The type size of fonts stresses the importance and emphasis of the intended map. Size is expressed in points through the American point system with 1 point equaling 1/72" of vertical height. Furthermore, points also show the spacing between letters, words and lines. A larger size implies more importance or a greater relative quantity; smaller denotes less importance or less quantity. For design purposes, text using a size of less than 6 point is difficult to read. On the contraire, text that is larger than 26 point is too cumbersome for a standard-size paper format. For titles, a font larger than 10 point generally allows for a good working title. Also, it is important to use at least a 2-point difference between type sizes to allow the audience to see subtle changes.

Placement

With all of the type in order and adequately designed, the final step is the correct placement of labels. Placement describes each feature and its subsequent label(s). For area features, it is important to curve and extend the spaces to properly fill in the areas enough that the audience can discern different areas. As a cartographic convention, labels are usually as horizontal as possible with no upside-down labels. For line features, it is useful to allow the label to conform to the line pattern. Similar to a river (e.g. geographic features), the label should flow around the edges along the line being careful not to have the letters too extended. For point patterns, the minor patterns to follow include keeping labels on/in their respective features (e.g. coastal cities with labels on the land and not ocean). The major pattern for points is the placement along the point itself. The most widely accepted pattern is to start at the center and work outward towards the northeast quadrant from the point. Many studies have been researched to address the correct strategy for the placements. The point feature cartographic label placement (PFCLP) problem offers the solutions when point boxes overlap. Many software features automatically choose label placements for the cartographer, but these are not always a fail-safe option. The use of good judgment and cartographic conventions are important to gain the best placement.

Cartographic Generalization

Cartographic generalization, or map generalization, is a method for deriving a smaller-scale map from a larger scale map or map data Whether done manually by a cartographer or by a computer or set of algorithms, generalization seeks to abstract spatial information at a high level of detail to information that can be rendered on a map at a lower level of detail. For example, we might have the outlines of all of the thousands of buildings in a region, but we wish to make a map of the whole city no more than a few inches wide. Instead of throwing out the building information, or trying to render it all at once, we could generalize the data into some sort of outline of the urbanized area of the region.

The cartographer has license to adjust the content within their maps to create a suitable and useful map that conveys spatial information, while striking the right balance between the map's purpose and the precise detail of the subject being mapped. Well generalized maps are those that emphasize the most important map elements while still representing the world in the most faithful and recognizable way.

Methods

There are many cartographic techniques that may fall into the broad category of generalization. Brief descriptions of some of the more common methods follow.

Selection

Map generalization is designed to reduce the complexities of the real world by strategically reducing ancillary and unnecessary details. One way that geospatial data can be reduced is through the selection process. The cartographer can select and retain certain elements that he/she deems the most necessary or appropriate. In this method, the most important elements stand out while lesser elements are left out entirely. For example, a directional map between two points may have lesser and un-traveled roadways omitted as not to confuse the map-reader. The selection of the most direct and uncomplicated route between the two points is the most important data, and the cartographer may choose to emphasize this.

Simplification

Generalization is not a process that only removes and selects data, but also a process that simplifies or abstracts it as well. Simplification is a technique where the general shapes of features are retained, while eliminating unnecessary detail. Generally, smaller scale maps have more simplified features than larger scale maps. One common line or polygon generalization technique is the Ramer–Douglas–Peucker algorithm.

Combination

Simplification also takes on other roles when considering the role of combination. Overall data reduction techniques can also mean that in addition to generalizing elements of particular features, features can also be combined when their separation is irrelevant to the map focus. A mountain chain may be isolated into several smaller ridges and peaks with intermittent forest in the natural environment, but shown as a continuous chain on the map, as determined by scale. The map reader has to, again remember, that because of scale limitations combined elements are not concise depictions of natural or manmade features.

Smoothing

Smoothing is also a process that the map maker can employ to reduce the angularity of line work. Smoothing is yet another way of simplifying the map features, but involves several other characteristics of generalization that lead into feature displacement and locational shifting. The purpose of smoothing is to exhibit linework in a much less complicated and a less visually jarring way. An example of smoothing would be for a jagged roadway, cut through a mountain, to be smoothed out so that the angular turns and transitions appear much more fluid and natural.

Enhancement

Enhancement is also a method that can be employed by the cartographer to illuminate specific elements that aid in map reading. As many of the aforementioned generalizing methods focus on the reduction and omission of detail, the enhancement method concentrates on the addition of detail. Enhancement can be used to show the true character of the feature being represented and is often used by the cartographer to highlight specific details about his or her specific knowledge, that would otherwise be left out. An example includes enhancing the detail about specific river rapids so that the map reader may know the facets of traversing the most difficult sections beforehand. Enhancement can be a valuable tool in aiding the map reader to elements that carry significant weight to the map's intent.

Displacement

Displacement can be employed when 2 objects are so close to each other that they would overlap at smaller scales. A common place where this would occur is the cities Brazzaville and Kinshasa on either side of the Congo river in Africa. They are both the capital city of their country and on overview maps they would be displayed with a slightly larger symbol than other cities. Depending on the scale of the map the symbols would overlap. By displacing both of them away from the river (and away from their true location) the symbol overlap can be avoided. Another common case is when a road and a railroad run parallel to each other.

GIS and Automated Generalization

As GIS developed from about the late 1960s onward, the need for automatic, algorithmic generalization techniques became clear. Ideally, agencies responsible for collecting and maintaining spatial data should try to keep only one canonical representation of a given feature, at the highest possible level of detail. That way there is only one record to update when that feature changes in the real world. From this large-scale data, it should ideally be possible, through automated generalization, to produce maps and other data products at any scale required. The alternative is to maintain separate databases each at the scale required for a given set of mapping projects, each of which requires attention when something changes in the real world.

Several broad approaches to generalization were developed around this time:

- The *representation-oriented* view focuses on the representation of data on different scales, which is related to the field of Multi-Representation Databases (MRDB).
- The *process-oriented* view focuses on the process of generalization.
- The *ladder-approach* is a stepwise generalization, in which each derived dataset is based on the other database of the next larger scale.
- The *star-approach* is the derived data on all scales is based on a single (large-scale) data base.

The 'Baltimore Phenomenon'

The Baltimore Phenomenon is the tendency for a city (or other object) to be omitted from maps due to space constraints while smaller cities are included on the same map simply because space is available to display them. This phenomenon gets its name from Baltimore, Maryland, which, despite its large population, is commonly omitted on maps of the United States because there is not enough space in the surrounding area of the map. Larger cities surrounding Baltimore take precedence. In contrast, much smaller cities in other geographic locations are included at the same scale because the level of competition for map space may not exist in that particular area.

Map Projection

Commonly, a map projection is a systematic transformation of the latitudes and longitudes of locations on the surface of a sphere or an ellipsoid into locations on a plane. Map projections are necessary for creating maps. All map projections distort the surface in some fashion. Depending on the purpose of the map, some distortions are acceptable and others are not; therefore, different map projections exist in order to preserve some properties of the sphere-like body at the expense of other properties. There is no limit to the number of possible map projections.[1]

More generally, the surfaces of planetary bodies can be mapped even if they are too irregular to be modeled well with a sphere or ellipsoid. Even more generally, projections are the subject of several pure mathematical fields, including differential geometry and projective geometry. However, "map projection" refers specifically to a cartographic projection.

Background

Maps can be more useful than globes in many situations: they are more compact and easier to store; they readily accommodate an enormous range of scales; they are viewed easily on computer displays; they can facilitate measuring properties of the terrain being mapped; they can show larger portions of the Earth's surface at once; and they are cheaper to produce and transport. These useful traits of maps motivate the development of map projections.

However, Carl Friedrich Gauss's Theorema Egregium proved that a sphere's surface cannot be represented on a plane without distortion. The same applies to other reference surfaces used as models for the Earth. Since any map projection is a representation of one of those surfaces on a plane, all map projections distort. Every distinct map projection distorts in a distinct way. The study of map projections is the characterization of these distortions.

Projection is not limited to perspective projections, such as those resulting from casting a shadow on a screen, or the rectilinear image produced by a pinhole camera on a flat film plate. Rather, any mathematical function transforming coordinates from the curved surface to the plane is a projection. Few projections in actual use are perspective.

For simplicity, most of this article assumes that the surface to be mapped is that of a sphere. In reality, the Earth and other large celestial bodies are generally better modeled as oblate spheroids, whereas small objects such as asteroids often have irregular shapes. These other surfaces can be mapped as well. Therefore, more generally, a map projection is any method of "flattening" into a plane a continuous curved surface.

Metric Properties of Maps

Map projections can be constructed to preserve at least one of these properties, though only in a limited way for most. Each projection preserves or compromises or approximates basic metric properties in different ways. The purpose of the map determines which projection should form the base for the map. Because many purposes exist for maps, many projections have been created to suit those purposes.

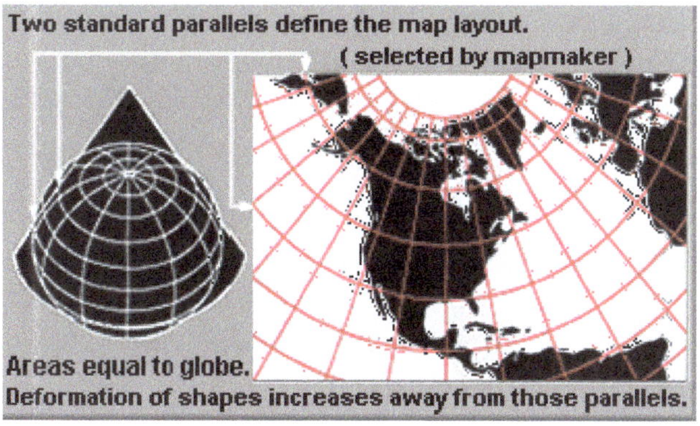

An Albers projection shows areas accurately, but distorts shapes.

Many properties can be measured on the Earth's surface independent of its geography. Some of these properties are:

- Area
- Shape
- Direction
- Bearing
- Distance
- Scale

Another consideration in the configuration of a projection is its compatibility with data sets to be used on the map. Data sets are geographic information; their collection depends on the chosen datum (model) of the Earth. Different datums assign slightly different coordinates to the same location, so in large scale maps, such as those from national mapping systems, it is important to match the datum to the projection. The slight differences in coordinate assignation between different datums is not a concern for world maps or other vast territories, where such differences get shrunk to imperceptibility.

Which Projection is Best?

The mathematics of projection do not permit any particular map projection to be "best" for everything. Something will always get distorted. Therefore, a diversity of projections exists to service the many uses of maps and their vast range of scales.

Modern national mapping systems typically employ a transverse Mercator or close variant for large-scale maps in order to preserve conformality and low variation in scale over small areas. For smaller-scale maps, such as those spanning continents or the entire world, many projections are in common use according to their fitness for the purpose.

Thematic maps normally require an equal area projection so that phenomena per unit area are shown in correct proportion. However, representing area ratios correctly necessarily distorts shapes more than many maps that are not equal-area. Hence reference maps of the world often appear on compromise projections instead. Due to distortions inherent in any map of the world, the choice of projection becomes largely one of aesthetics.

The Mercator projection, developed for navigational purposes, has often been used in world maps where other projections would have been more appropriate. This problem has long been recognized even outside professional circles. For example, a 1943 *New York Times* editorial states:

The time has come to discard [the Mercator] for something that represents the continents and directions less deceptively... Although its usage... has diminished... it is still highly popular as a wall map apparently in part because, as a rectangular map, it fills a rectangular wall space with more map, and clearly because its familiarity breeds more popularity.

A controversy in the 1980s over the Peters map motivated the American Cartographic Association (now Cartography and Geographic Information Society) to produce a series of booklets (including *Which Map Is Best*) designed to educate the public about map projections and distortion in maps. In 1989 and 1990, after some internal debate, seven North American geographic organizations adopted a resolution recommending against using any rectangular projection (including Mercator and Gall–Peters) for reference maps of the world.

Distortion

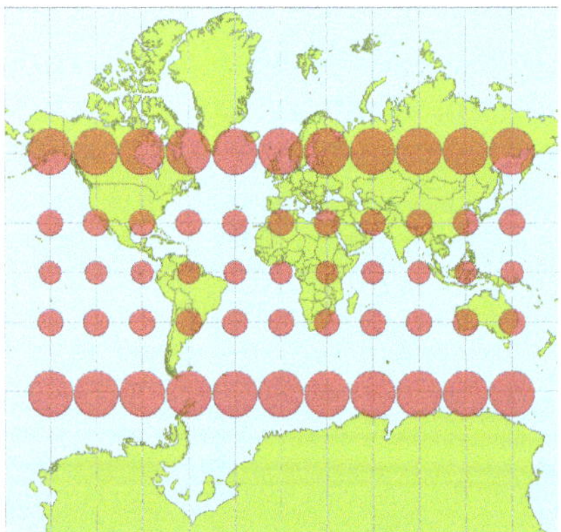

Tissot's Indicatrices on the Mercator projection

The classical way of showing the distortion inherent in a projection is to use Tissot's indicatrix. For a given point, using the scale factor h along the meridian, the scale factor k along the parallel, and the angle θ' between them, Nicolas Tissot described how to construct an ellipse that characterizes the amount and orientation of the components of distortion.[:147–149] By spacing the ellipses regularly along the meridians and parallels, the network of indicatrices shows how distortion varies across the map.

Construction of a Map Projection

The creation of a map projection involves two steps:

1. Selection of a model for the shape of the Earth or planetary body (usually choosing between a sphere or ellipsoid). Because the Earth's actual shape is irregular, information is lost in this step.
2. Transformation of geographic coordinates (longitude and latitude) to Cartesian (x,y) or polar plane coordinates. Cartesian coordinates normally have a simple relation to eastings and northings defined on a grid superimposed on the projection.

Some of the simplest map projections are literal projections, as obtained by placing a light source at some definite point relative to the globe and projecting its features onto a specified surface. This is not the case for most projections, which are defined **only** in terms of mathematical formulae that have no direct geometric interpretation.

Choosing a Projection Surface

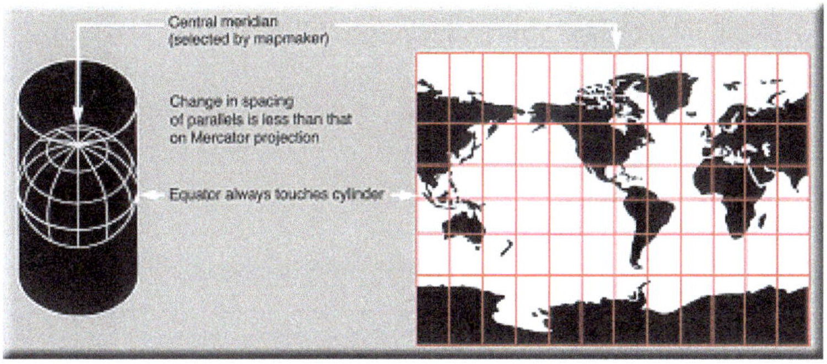

A Miller cylindrical projection maps the globe onto a cylinder.

A surface that can be unfolded or unrolled into a plane or sheet without stretching, tearing or shrinking is called a *developable surface*. The cylinder, cone and the plane are all developable surfaces. The sphere and ellipsoid do not have developable surfaces, so any projection of them onto a plane will have to distort the image. (To compare, one cannot flatten an orange peel without tearing and warping it.)

One way of describing a projection is first to project from the Earth's surface to a developable surface such as a cylinder or cone, and then to unroll the surface into a plane. While the first step inevitably distorts some properties of the globe, the developable surface can then be unfolded without further distortion.

Aspect of the Projection

Once a choice is made between projecting onto a cylinder, cone, or plane, the aspect of the shape must be specified. The aspect describes how the developable surface is placed relative to the globe: it may be *normal* (such that the surface's axis of symmetry coincides with the Earth's axis), *transverse* (at right angles to the Earth's axis) or *oblique* (any angle in between).

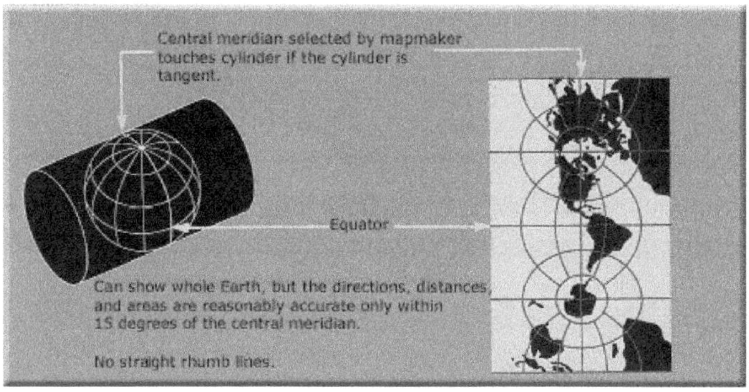

This transverse Mercator projection is mathematically the same as a standard Mercator, but oriented around a different axis.

Notable Lines

The developable surface may also be either *tangent* or *secant* to the sphere or ellipsoid. Tangent means the surface touches but does not slice through the globe; secant means the surface does slice through the globe. Moving the developable surface away from contact with the globe never preserves or optimizes metric properties, so that possibility is not discussed further here.

Tangent and secant lines (*standard lines*) are represented undistorted. If these lines are a parallel of latitude, as in conical projections, it is called a *standard parallel*. The *central meridian* is the meridian to which the globe is rotated before projecting. The central meridian (usually written λ_0) and a parallel of origin (usually written φ_0) are often used to define the origin of the map projection.

Scale

A globe is the only way to represent the earth with constant scale throughout the entire map in all directions. A map cannot achieve that property for any area, no matter how small. It can, however, achieve constant scale along specific lines.

Some possible properties are:

- The scale depends on location, but not on direction. This is equivalent to preservation of angles, the defining characteristic of a conformal map.
- Scale is constant along any parallel in the direction of the parallel. This applies for any cylindrical or pseudocylindrical projection in normal aspect.
- Combination of the above: the scale depends on latitude only, not on longitude or direction. This applies for the Mercator projection in normal aspect.

- Scale is constant along all straight lines radiating from a particular geographic location. This is the defining characteristic of an equidistant projection such as the Azimuthal equidistant projection. There are also projections (Maurer, Close) where true distances from *two* points are preserved.

Choosing a Model for the Shape of the Body

Projection construction is also affected by how the shape of the Earth or planetary body is approximated. In the following section on projection categories, the earth is taken as a sphere in order to simplify the discussion. However, the Earth's actual shape is closer to an oblate ellipsoid. Whether spherical or ellipsoidal, the principles discussed hold without loss of generality.

Selecting a model for a shape of the Earth involves choosing between the advantages and disadvantages of a sphere versus an ellipsoid. Spherical models are useful for small-scale maps such as world atlases and globes, since the error at that scale is not usually noticeable or important enough to justify using the more complicated ellipsoid. The ellipsoidal model is commonly used to construct topographic maps and for other large- and medium-scale maps that need to accurately depict the land surface. Auxiliary latitudes are often employed in projecting the ellipsoid.

A third model is the geoid, a more complex and accurate representation of Earth's shape coincident with what mean sea level would be if there were no winds, tides, or land. Compared to the best fitting ellipsoid, a geoidal model would change the characterization of important properties such as distance, conformality and equivalence. Therefore, in geoidal projections that preserve such properties, the mapped graticule would deviate from a mapped ellipsoid's graticule. Normally the geoid is not used as an Earth model for projections, however, because Earth's shape is very regular, with the undulation of the geoid amounting to less than 100 m from the ellipsoidal model out of the 6.3 million m Earth radius. For irregular planetary bodies such as asteroids, however, sometimes models analogous to the geoid are used to project maps from.

Classification

A fundamental projection classification is based on the type of projection surface onto which the globe is conceptually projected. The projections are described in terms of placing a gigantic surface in contact with the earth, followed by an implied scaling operation. These surfaces are cylindrical (e.g. Mercator), conic (e.g. Albers), or azimuthal or plane (e.g. stereographic). Many mathematical projections, however, do not neatly fit into any of these three conceptual projection methods. Hence other peer categories have been described in the literature, such as pseudoconic, pseudocylindrical, pseudoazimuthal, retroazimuthal, and polyconic.

Another way to classify projections is according to properties of the model they preserve. Some of the more common categories are:

- Preserving direction (*azimuthal or zenithal*), a trait possible only from one or two points to every other point
- Preserving shape locally (*conformal* or *orthomorphic*)
- Preserving area (*equal-area* or *equiareal* or *equivalent* or *authalic*)
- Preserving distance (*equidistant*), a trait possible only between one or two points and every other point
- Preserving shortest route, a trait preserved only by the gnomonic projection

Because the sphere is not a developable surface, it is impossible to construct a map projection that is both equal-area and conformal.

Projections by Surface

The three developable surfaces (plane, cylinder, cone) provide useful models for understanding, describing, and developing map projections. However, these models are limited in two fundamental ways. For one thing, most world projections in actual use do not fall into any of those categories. For another thing, even most projections that do fall into those categories are not naturally attainable through physical projection. As L.P. Lee notes,

No reference has been made in the above definitions to cylinders, cones or planes. The projections are termed cylindric or conic because they can be regarded as developed on a cylinder or a cone, as the case may be, but it is as well to dispense with picturing cylinders and cones, since they have given rise to much misunderstanding. Particularly is this so with regard to the conic projections with two standard parallels: they may be regarded as developed on cones, but they are cones which bear no simple relationship to the sphere. In reality, cylinders and cones provide us with convenient descriptive terms, but little else.

Lee's objection refers to the way the terms *cylindrical, conic*, and *planar* (azimuthal) have been abstracted in the field of map projections. If maps were projected as in light shining through a globe onto a developable surface, then the spacing of parallels would follow a very limited set of possibilities. Such a cylindrical projection (for example) is one which:

1. Is rectangular;
2. Has straight vertical meridians, spaced evenly;
3. Has straight parallels symmetrically placed about the equator;
4. Has parallels constrained to where they fall when light shines through the globe onto the cylinder, with the light source someplace along the line formed by the intersection of the prime meridian with the equator, and the center of the sphere.

(If you rotate the globe before projecting then the parallels and meridians will not necessarily still be straight lines. Rotations are normally ignored for the purpose of classification.)

Where the light source emanates along the line described in this last constraint is what yields the differences between the various "natural" cylindrical projections. But the term *cylindrical* as used in the field of map projections relaxes the last constraint entirely. Instead the parallels can be placed according to any algorithm the designer has decided suits the needs of the map. The famous Mercator projection is one in which the placement of parallels does not arise by "projection"; instead parallels are placed how they need to be in order to satisfy the property that a course of constant bearing is always plotted as a straight line.

Cylindrical

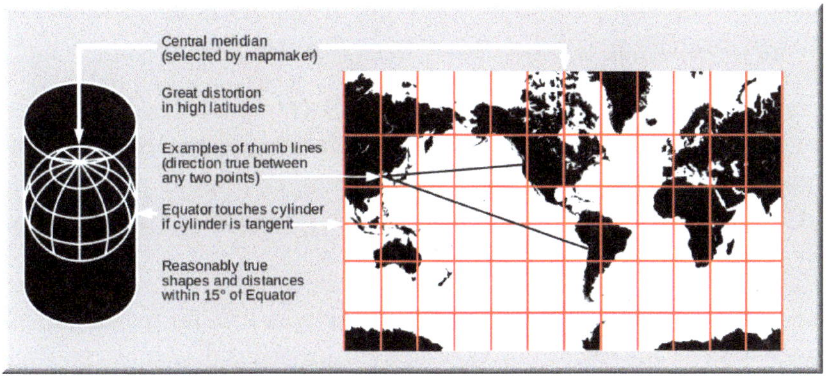

The Mercator projection shows rhumbs as straight lines. A rhumb is a course of constant bearing. Bearing is the compass direction of movement.

The term "normal cylindrical projection" is used to refer to any projection in which meridians are mapped to equally spaced vertical lines and circles of latitude (parallels) are mapped to horizontal lines.

The mapping of meridians to vertical lines can be visualized by imagining a cylinder whose axis coincides with the Earth's axis of rotation. This cylinder is wrapped around the Earth, projected onto, and then unrolled.

By the geometry of their construction, cylindrical projections stretch distances east-west. The amount of stretch is the same at any chosen latitude on all cylindrical projections, and is given by the secant of the latitude as a multiple of the equator's scale. The various cylindrical projections are distinguished from each other solely by their north-south stretching (where latitude is given by φ):

- North-south stretching equals east-west stretching (sec φ): The east-west scale matches the north-south scale: conformal cylindrical or Mercator; this distorts areas excessively in high latitudes.

- North-south stretching grows with latitude faster than east-west stretching ($\sec^2 \varphi$): The cylindric perspective (or central cylindrical) projection; unsuitable because distortion is even worse than in the Mercator projection.
- North-south stretching grows with latitude, but less quickly than the east-west stretching: such as the Miller cylindrical projection ($\sec 4\varphi/5$).
- North-south distances neither stretched nor compressed (1): equirectangular projection or "plate carrée".
- North-south compression equals the cosine of the latitude (the reciprocal of east-west stretching): equal-area cylindrical. This projection has many named specializations differing only in the scaling constant, such as the Gall–Peters or Gall orthographic (undistorted at the 45° parallels), Behrmann (undistorted at the 30° parallels), and Lambert cylindrical equal-area (undistorted at the equator). Since this projection scales north-south distances by the reciprocal of east-west stretching, it preserves area at the expense of shapes.

In the first case (Mercator), the east-west scale always equals the north-south scale. In the second case (central cylindrical), the north-south scale exceeds the east-west scale everywhere away from the equator. Each remaining case has a pair of secant lines—a pair of identical latitudes of opposite sign (or else the equator) at which the east-west scale matches the north-south-scale.

Normal cylindrical projections map the whole Earth as a finite rectangle, except in the first two cases, where the rectangle stretches infinitely tall while retaining constant width.

Pseudocylindrical

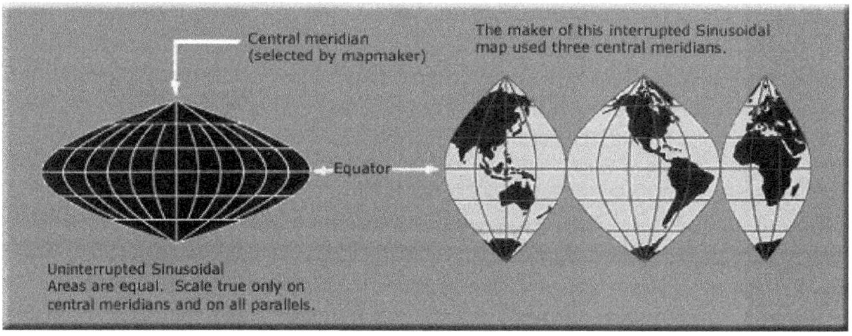

A sinusoidal projection shows relative sizes accurately, but grossly distorts shapes. Distortion can be reduced by "interrupting" the map.

Pseudocylindrical projections represent the *central* meridian as a straight line segment. Other meridians are longer than the central meridian and bow outward away from the central meridian. Pseudocylindrical projections map parallels as straight lines. Along parallels, each point from the surface is mapped at a distance from the central meridian that is proportional to its difference in longitude from the central meridian. On a pseudocylindrical map, any point further from the equator than some other point has a higher latitude than the other point, preserving north-south relationships. This trait is useful when illustrating phenomena that depend on latitude, such as climate. Examples of pseudocylindrical projections include:

- Sinusoidal, which was the first pseudocylindrical projection developed. Vertical scale and horizontal scale are the same throughout, resulting in an equal-area map. On the map, as in reality, the length of each parallel is proportional to the cosine of the latitude. Thus the shape of the map for the whole earth is the region between two symmetric rotated cosine curves. The true distance between two points on the same meridian corresponds to the distance on the map between the two parallels, which is smaller than the distance between the two points on the map. The distance between two points on the same parallel is true. The area of any region is true.

- Collignon projection, which in its most common forms represents each meridian as two straight line segments, one from each pole to the equator.

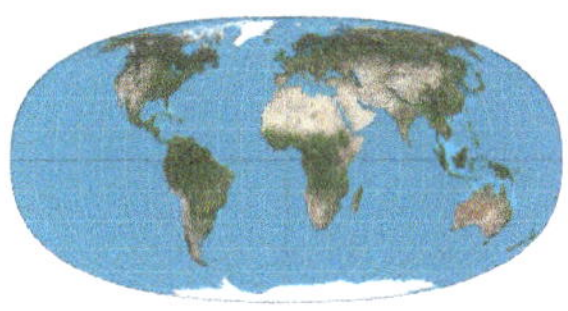

• Tobler hyperelliptical

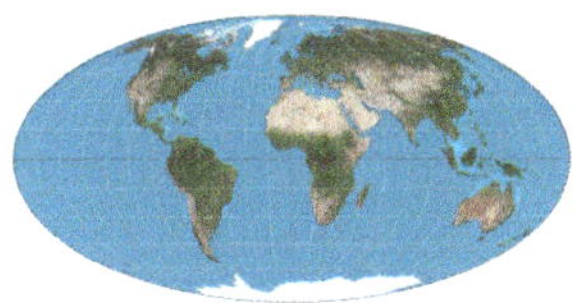

• Mollweide

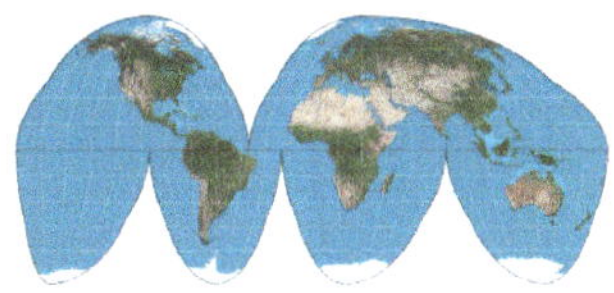

• Goode homolosine

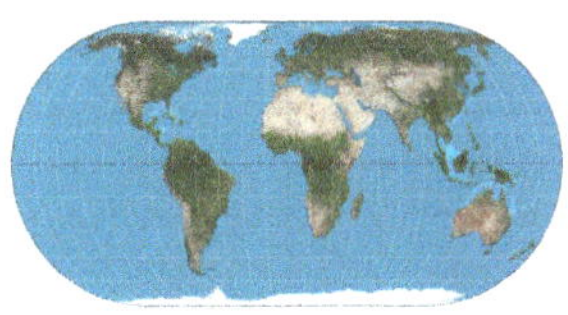

• Eckert IV

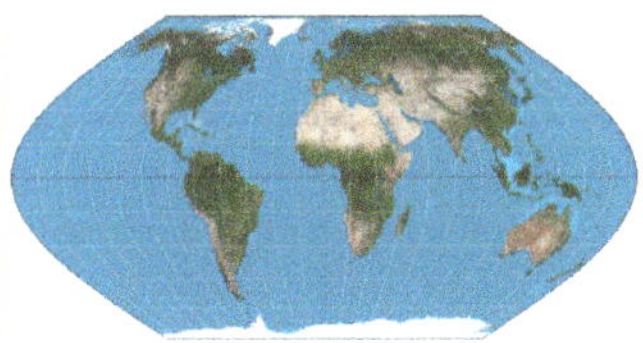

• Eckert VI

• Kavrayskiy VII

Hybrid

The HEALPix projection combines an equal-area cylindrical projection in equatorial regions with the Collignon projection in polar areas.

Conic

The term "conic projection" is used to refer to any projection in which meridians are mapped to equally spaced lines radiating out from the apex and circles of latitude (parallels) are mapped to circular arcs centered on the apex.

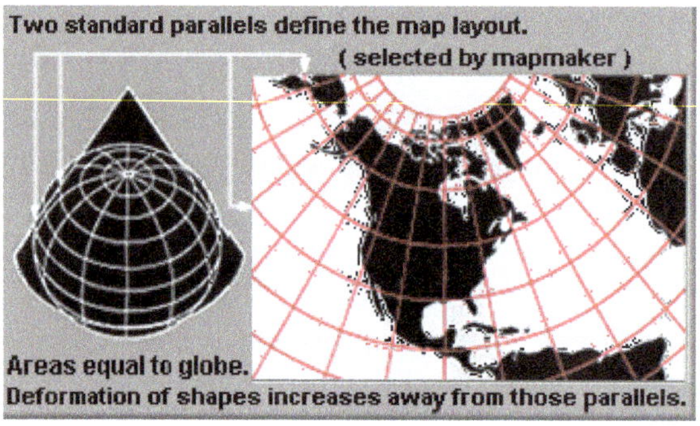

Albers conic.

When making a conic map, the map maker arbitrarily picks two standard parallels. Those standard parallels may be visualized as secant lines where the cone intersects the globe—or, if the map maker chooses the same parallel twice, as the tangent line where the cone is tangent to the globe. The resulting conic map has low distortion in scale, shape, and area near those standard parallels. Distances along the parallels to the north of both standard parallels or to the south of both standard parallels are stretched; distances along parallels between the standard parallels are compressed. When a single standard parallel is used, distances along all other parallels are stretched.

The most popular conic maps include:

- Equidistant conic, which keeps parallels evenly spaced along the meridians to preserve a constant distance scale along each meridian, typically the same or similar scale as along the standard parallels.
- Albers conic, which adjusts the north-south distance between non-standard parallels to compensate for the east-west stretching or compression, giving an equal-area map.
- Lambert conformal conic, which adjusts the north-south distance between non-standard parallels to equal the east-west stretching, giving a conformal map.

Pseudoconic

- Bonne
- Werner cordiform, upon which distances are correct from one pole, as well as along all parallels.
- Continuous American polyconic

Azimuthal (Projections Onto a Plane)

Azimuthal projections have the property that directions from a central point are preserved and therefore great circles through the central point are represented by straight lines on the map. Usually these projections also have radial symmetry in the scales and hence in the distortions: map distances from the central point are computed by a function $r(d)$ of the true distance d, independent of the angle; correspondingly, circles with the central point as center are mapped into circles which have as center the central point on the map.

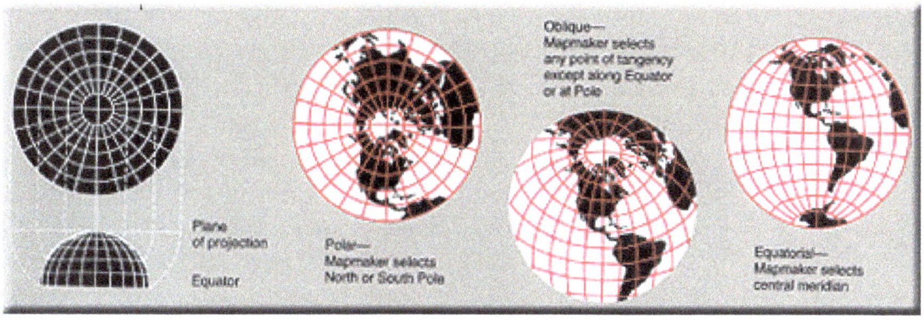

An azimuthal equidistant projection shows distances and directions accurately from the center point, but distorts shapes and sizes elsewhere.

The mapping of radial lines can be visualized by imagining a plane tangent to the Earth, with the central point as tangent point.

The radial scale is $r'(d)$ and the transverse scale $r(d)/(R \sin d/R)$ where R is the radius of the Earth.

Some azimuthal projections are true perspective projections; that is, they can be constructed mechanically, projecting the surface of the Earth by extending lines from a point of perspective (along an infinite line through the tangent point and the tangent point's antipode) onto the plane:

- The gnomonic projection displays great circles as straight lines. Can be constructed by using a point of perspective at the center of the Earth. $r(d) = c \tan d/R$; so that even just a hemisphere is already infinite in extent.
- The General Perspective projection can be constructed by using a point of perspective outside the earth. Photographs of Earth (such as those from the International Space Station) give this perspective.
- The orthographic projection maps each point on the earth to the closest point on the plane. Can be constructed from a point of perspective an infinite distance from the tangent point; $r(d) = c \sin d/R$. Can display up to a hemisphere on a finite circle. Photographs of Earth from far enough away, such as the Moon, give this perspective.
- The azimuthal conformal projection, also known as the stereographic projection, can be constructed by using the tangent point's antipode as the point of perspective. $r(d) = c \tan d/2R$; the scale is $c/(2R \cos^2 d/2R)$. Can display nearly

the entire sphere's surface on a finite circle. The sphere's full surface requires an infinite map.

Other azimuthal projections are not true perspective projections:

- Azimuthal equidistant: $r(d) = cd$; it is used by amateur radio operators to know the direction to point their antennas toward a point and see the distance to it. Distance from the tangent point on the map is proportional to surface distance on the earth (; for the case where the tangent point is the North Pole, see the flag of the United Nations)

- Lambert azimuthal equal-area. Distance from the tangent point on the map is proportional to straight-line distance through the earth: $r(d) = c \sin d/2R$

- Logarithmic azimuthal is constructed so that each point's distance from the center of the map is the logarithm of its distance from the tangent point on the Earth. $r(d) = c \ln d/d_0$); locations closer than at a distance equal to the constant d_0 are not shown.

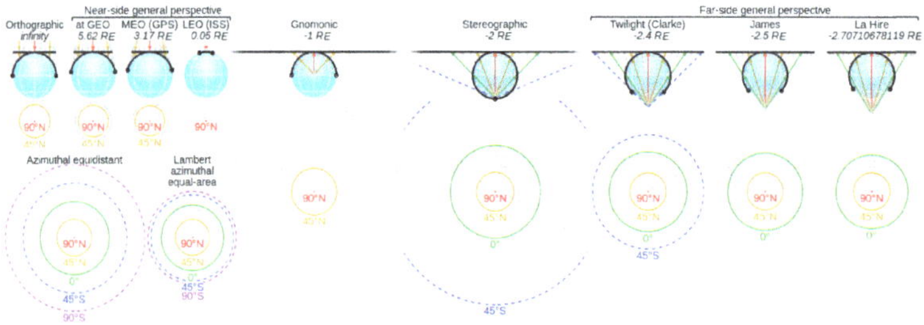

Comparison of some azimuthal projections centred on 90° N at the same scale, ordered by projection altitude in Earth radii. (click for detail)

Projections by Preservation of a Metric Property

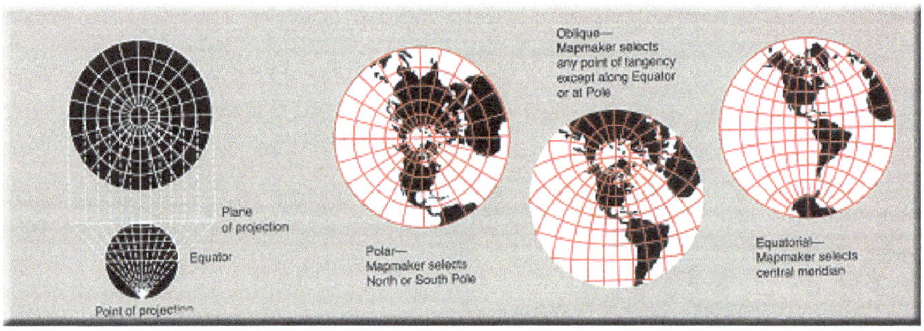

A stereographic projection is conformal and perspective but not equal area or equidistant.

Conformal

Conformal, or orthomorphic, map projections preserve angles locally, implying that they map infinitesimal circles of constant size anywhere on the Earth to infinitesimal circles of varying sizes on the map. In contrast, mappings that are not conformal distort most such small circles into ellipses of distortion. An important consequence of conformality is that relative angles at each point of the map are correct, and the local scale (although varying throughout the map) in every direction around any one point is constant. These are some conformal projections:

- Mercator: Rhumb lines are represented by straight segments
- Transverse Mercator
- Stereographic: Any circle of a sphere, great and small, maps to a circle or straight line.
- Roussilhe
- Lambert conformal conic
- Peirce quincuncial projection
- Adams hemisphere-in-a-square projection
- Guyou hemisphere-in-a-square projection

Equal-area

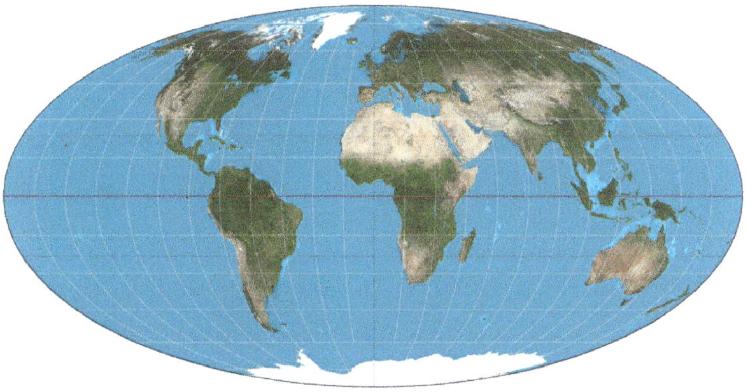

The equal-area Mollweide projection

Equal-area maps preserve area measure, generally distorting shapes in order to do that. Equal-area maps are also called *equivalent* or *authalic*. These are some projections that preserve area:

- Albers conic
- Bonne
- Bottomley
- Collignon
- Cylindrical equal-area

- Eckert II, IV and VI
- Gall orthographic (also known as Gall–Peters, or Peters, projection)
- Goode's homolosine
- Hammer
- Hobo–Dyer
- Lambert azimuthal equal-area
- Lambert cylindrical equal-area
- Mollweide
- Sinusoidal
- Snyder's equal-area polyhedral projection, used for geodesic grids.
- Tobler hyperelliptical
- Werner

Equidistant

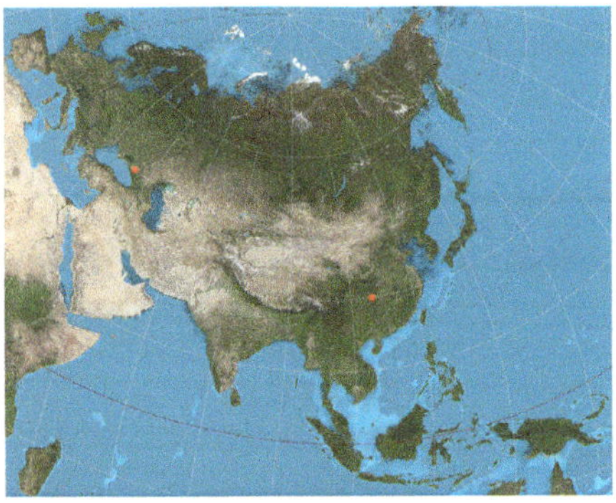

A two-point equidistant projection of Eurasia

These are some projections that preserve distance from some standard point or line:

- Equirectangular—distances along meridians are conserved
- Plate carrée—an Equirectangular projection centered at the equator
- Azimuthal equidistant—distances along great circles radiating from centre are conserved
- Equidistant conic
- Sinusoidal—distances along parallels are conserved
- Werner cordiform distances from the North Pole are correct as are the curved distance on parallels

- Soldner
- Two-point equidistant: two "control points" are arbitrarily chosen by the map maker. Distance from any point on the map to each control point is proportional to surface distance on the earth.

Gnomonic

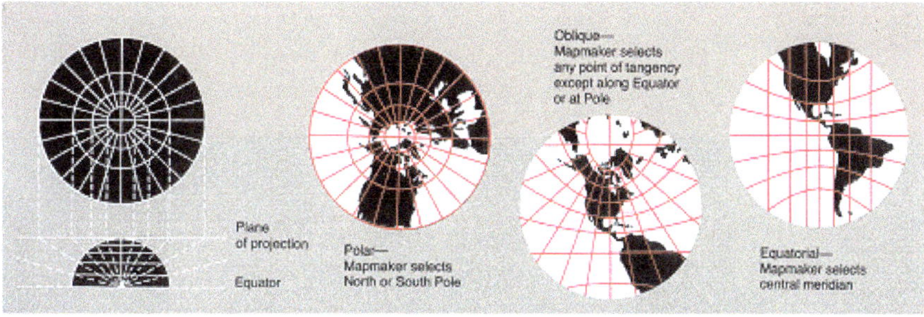

The Gnomonic projection is thought to be the oldest map projection, developed by Thales in the 6th century BC

Great circles are displayed as straight lines:

- Gnomonic projection

Retroazimuthal

Direction to a fixed location B (the bearing at the starting location A of the shortest route) corresponds to the direction on the map from A to B:

- Littrow—the only conformal retroazimuthal projection
- Hammer retroazimuthal—also preserves distance from the central point
- Craig retroazimuthal *aka* Mecca or Qibla—also has vertical meridians

Compromise Projections

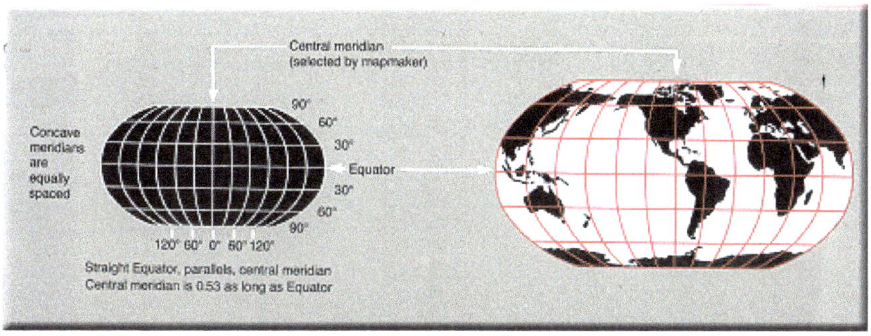

The Robinson projection was adopted by *National Geographic* magazine in 1988 but abandoned by them in about 1997 for the Winkel tripel.

Compromise projections give up the idea of perfectly preserving metric properties, seeking instead to strike a balance between distortions, or to simply make things "look right". Most of these types of projections distort shape in the polar regions more than at the equator. These are some compromise projections:

- Robinson
- van der Grinten
- Miller cylindrical
- Winkel Tripel
- Buckminster Fuller's Dymaxion
- B. J. S. Cahill's Butterfly Map
- Kavrayskiy VII projection
- Wagner VI projection
- Chamberlin trimetric
- Oronce Finé's cordiform

Contour Line

A contour line (also isoline, isopleth, or isarithm) of a function of two variables is a curve along which the function has a constant value. It is a cross-section of the three-dimensional graph of the function f(x, y) parallel to the x, y plane. In cartography, a contour line (often just called a "contour") joins points of equal elevation (height) above a given level, such as mean sea level. A contour map is a map illustrated with contour lines, for example a topographic map, which thus shows valleys and hills, and the steepness of slopes. The contour interval of a contour map is the difference in elevation between successive contour lines.

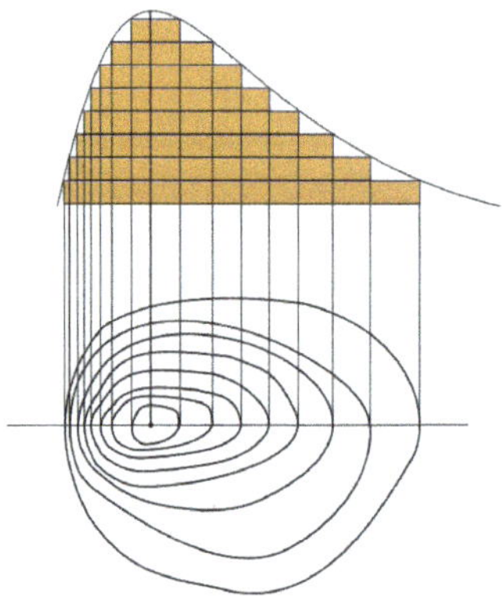

The bottom part of the diagram shows some contour lines with a straight line running through the location of the maximum value. The curve at the top represents the values along that straight line.

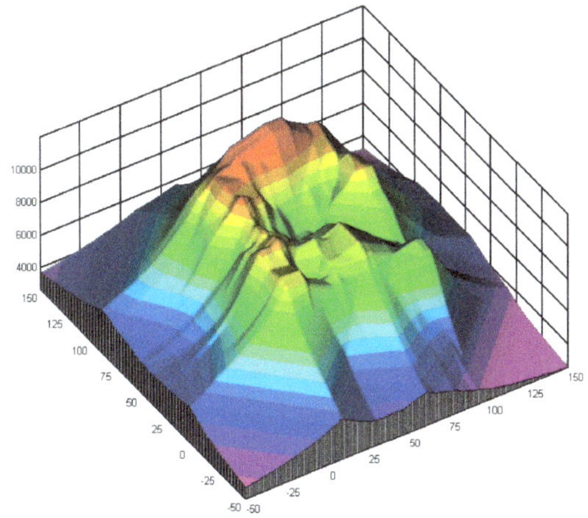

A three-dimensional surface, whose contour graph is below.

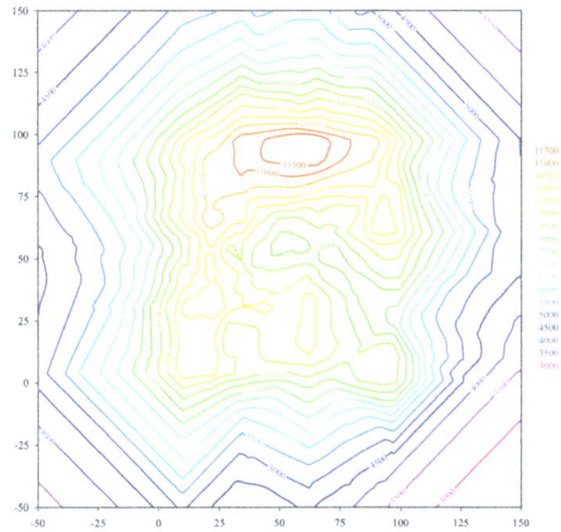

A two-dimensional contour graph of the three-dimensional surface in the above picture.

More generally, a contour line for a function of two variables is a curve connecting points where the function has the same particular value. The gradient of the function is always perpendicular to the contour lines. When the lines are close together the magnitude of the gradient is large: the variation is steep. A level set is a generalization of a contour line for functions of any number of variables.

Contour lines are curved, straight or a mixture of both lines on a map describing the intersection of a real or hypothetical surface with one or more horizontal planes. The configuration of these contours allows map readers to infer relative gradient of a parameter and estimate that parameter at specific places. Contour lines may be either traced on a visible three-dimensional model of the surface, as when a photogrammetrist viewing a stereo-model plots elevation contours, or interpolated from estimated surface elevations, as when a computer program threads contours through a network of observation points of area centroids. In the latter case, the method of interpolation affects the reliability of individual isolines and their portrayal of slope, pits and peaks.

Types

Contour lines are often given specific names beginning "iso-" according to the nature of the variable being mapped, although in many usages the phrase "contour line" is most commonly used. Specific names are most common in meteorology, where multiple maps with different variables may be viewed simultaneously. The prefix "iso-" can be replaced with "isallo-" to specify a contour line connecting points where a variable changes at the same *rate* during a given time period.

The words *isoline* and *isarithm* are general terms covering all types of contour line. The word *isogram* (γράμμα *gramma* "writing or drawing") was proposed by Francis Galton in 1889 as a convenient generic designation for lines indicating equality of some physical condition or quantity; but it commonly refers to a word without a repeated letter.

An isogon (from γωνία or *gonia*, meaning 'angle') is a contour line for a variable which measures direction. In meteorology and in geomagnetics, the term *isogon* has specific meanings which are described below. An isocline (from κλίνειν or *klinein*, meaning 'to lean or slope') is a line joining points with equal slope. In population dynamics and in geomagnetics, the terms *isocline* and *isoclinic line* have specific meanings which are described below.

Equidistants (Isodistances)

Equidistant is a line of equal distance from a given point, line, polyline.

Isopleths

In geography, the word *isopleth* (from *plethos*, meaning 'quantity') is used for contour lines that depict a variable which cannot be measured at a point, but which instead must be calculated from data collected over an area. An example is population density, which can be calculated by dividing the population of a census district by the surface area of that district. Each calculated value is presumed to be the value of the variable at the centre of the area, and isopleths can then be drawn by a process of interpolation. The idea of an isopleth map can be compared with that of a choropleth map.

In meteorology, the word *isopleth* is used for any type of contour line.

Meteorology

Meteorological contour lines are based on generalization from the point data received from weather stations. Weather stations are seldom exactly positioned at a contour line (when they are, this indicates a measurement precisely equal to the value of the contour). Instead, lines are drawn to best approximate the locations of exact values, based on the scattered information points available.

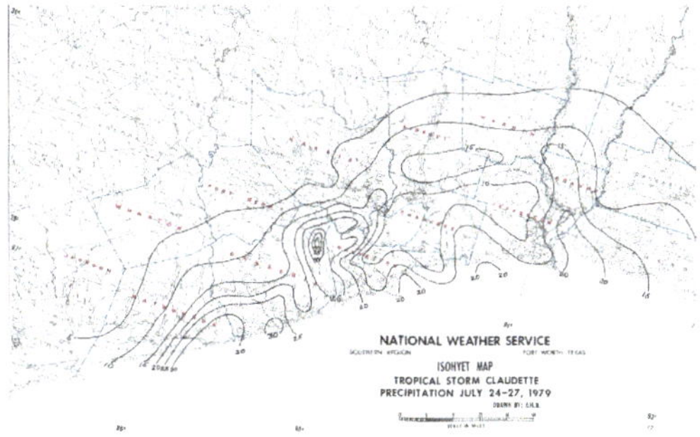

Isohyetal map of precipitation

Meteorological contour maps may present collected data such as actual air pressure at a given time, or generalized data such as average pressure over a period of time, or forecast data such as predicted air pressure at some point in the future

Thermodynamic diagrams use multiple overlapping contour sets (including isobars and isotherms) to present a picture of the major thermodynamic factors in a weather system.

Barometric Pressure

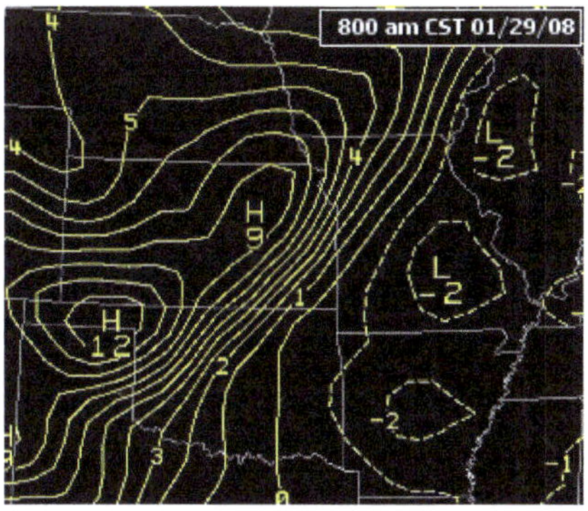

Loop showing the motion of a cold front by the movement of isallobars

An isobar (from βάρος or *baros*, meaning 'weight') is a line of equal or constant pressure on a graph, plot, or map; an isopleth or contour line of pressure. More accurately, isobars are lines drawn on a map joining places of equal average atmospheric pressure reduced to sea level for a specified period of time. In meteorology, the barometric pressures shown are reduced to sea level, not the surface pressures at the map locations. The distribution of isobars is closely related to the magnitude and direction of the wind field, and can be used to predict future weather patterns. Isobars are commonly used in television weather reporting.

Isallobars are lines joining points of equal pressure change during a specific time interval. These can be divided into *anallobars*, lines joining points of equal pressure increase during a specific time interval, and *katallobars*, lines joining points of equal pressure decrease. In general, weather systems move along an axis joining high and low isallobaric centers. Isallobaric gradients are important components of the wind as they increase or decrease the geostrophic wind.

An isopycnal is a line of constant density. An isoheight or isohypse is a line of constant geopotential height on a constant pressure surface chart.

Temperature and Related Subjects

The 10 °C (50 °F) mean isotherm in July, marked by the red line, is commonly used to define the border of the Arctic region

An isotherm (from θέρμη or *thermē*, meaning 'heat') is a line that connects points on a map that have the same temperature. Therefore, all points through which an isotherm passes have the same or equal temperatures at the time indicated. An isotherm at 0 °C is called the freezing level. The term was coined by the Prussian geographer and naturalist Alexander von Humboldt, who as part of his research into the geographical distribution of plants published the first map of isotherms in Paris, in 1817.

An isogeotherm is a line of equal mean annual temperature. An isocheim is a line of equal mean winter temperature, and an isothere is a line of equal mean summer temperature.

An isohel (from *helios*, meaning 'Sun') is a line of equal or constant solar radiation.

Rainfall and Air Moisture

An isohyet or isohyetal line (from *huetos*, meaning 'rain') is a line joining points of equal rainfall on a map in a given period . A map with isohyets is called an isohyetal map.

An isohume is a line of constant relative humidity, while a isodrosotherm (from δρόσος or drosos, meaning 'dew', and θέρμη or therme, meaning 'heat') is a line of equal or constant dew point.

An isoneph is a line indicating equal cloud cover.

An isochalaz is a line of constant frequency of hail storms, and an isobront is a line drawn through geographical points at which a given phase of thunderstorm activity occurred simultaneously.

Snow cover is frequently shown as a contour-line map.

Wind

An isotach (from ταχύς or *tachus*, meaning 'fast') is a line joining points with constant wind speed. In meteorology, the term isogon refers to a line of constant wind direction.

Freeze and Thaw

An isopectic line denotes equal dates of ice formation each winter, and an **isotac** denotes equal dates of thawing.

Physical Geography and Oceanography

Elevation and Depth

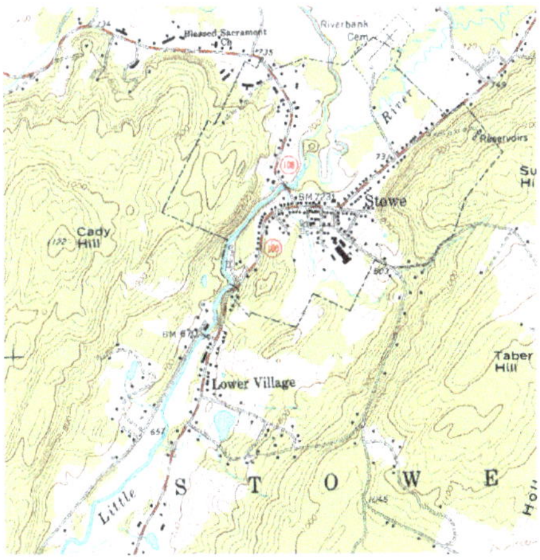

Topographic map of Stowe, Vermont. The brown contour lines represent the elevation. The contour interval is 20 feet.

Contours are one of several common methods used to denote elevation or altitude and depth on maps. From these contours, a sense of the general terrain can be determined. They are used at a variety of scales, from large-scale engineering drawings and architectural plans, through topographic maps and bathymetric charts, up to continental-scale maps.

"Contour line" is the most common usage in cartography, but isobath for underwater depths on bathymetric maps and isohypse for elevations are also used.

In cartography, the contour interval is the elevation difference between adjacent contour lines. The contour interval should be the same over a single map. When calculated as a ratio against the map scale, a sense of the hilliness of the terrain can be derived.

Interpretation

There are several rules to note when interpreting terrain contour lines:

- The rule of V's: sharp-pointed vees usually are in stream valleys, with the drainage channel passing through the point of the vee, with the vee pointing upstream. This is a consequence of erosion.
- The rule of O's: closed loops are normally uphill on the inside and downhill on the outside, and the innermost loop is the highest area. If a loop instead represents a depression, some maps note this by short lines radiating from the inside of the loop, called "hachures".
- Spacing of contours: close contours indicate a steep slope; distant contours a shallow slope. Two or more contour lines merging indicates a cliff. By counting the number of contours that cross a segment of a stream, you can approximate the stream gradient.

Of course, to determine differences in elevation between two points, the contour interval, or distance in altitude between two adjacent contour lines, must be known, and this is given at the bottom of the map. Usually contour intervals are consistent throughout a map, but there are exceptions. Sometimes intermediate contours are present in flatter areas; these can be dashed or dotted lines at half the noted contour interval. When contours are used with hypsometric tints on a small-scale map that includes mountains and flatter low-lying areas, it is common to have smaller intervals at lower elevations so that detail is shown in all areas. Conversely, for an island which consists of a plateau surrounded by steep cliffs, it is possible to use smaller intervals as the height increases.

Electrostatics

An isopotential map is a measure of electrostatic potential in space, often depicted in two dimensions with the electostatic charges inducing that electric potential. The term equipotential line or isopotential line refers to a curve of constant electric potential. Whether crossing an equipotential line represents ascending or descending the potential is inferred from the labels on the charges. In three dimensions, equipotential surfaces may be depicted with a two dimensional cross-section, showing equipotential lines at the intersection of the surfaces and the cross-section.

The general mathematical term level set is often used to describe the full collection of points having a particular potential, especially in higher dimensional space.

Magnetism

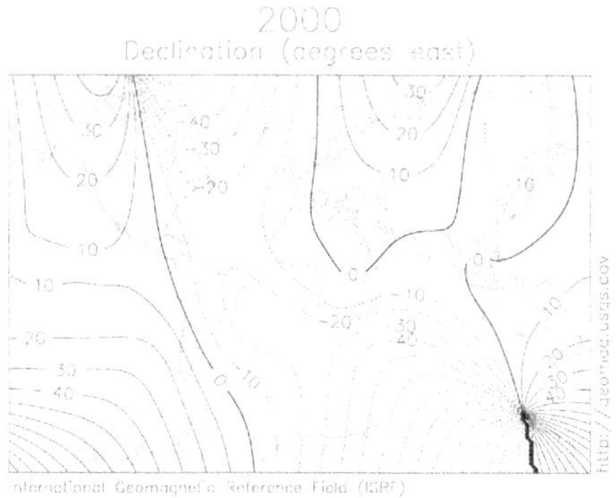

Isogonic lines for the year 2000. The agonic lines are thicker and labeled with “0”.

In the study of the Earth’s magnetic field, the term isogon or isogonic line refers to a line of constant magnetic declination, the variation of magnetic north from geographic north. An agonic line is drawn through points of zero magnetic declination. An isoporic line refers to a line of constant annual variation of magnetic declination .

An isoclinic line connects points of equal magnetic dip, and an aclinic line is the isoclinic line of magnetic dip zero.

An isodynamic line (from δύναμις or *dynamis* meaning ‘power’) connects points with the same intensity of magnetic force.

Oceanography

Besides ocean depth, oceanographers use contour to describe diffuse variable phenomena much as meteorologists do with atmospheric phenomena. In particular, isobathytherms are lines showing depths of water with equal temperature, isohalines show lines of equal ocean salinity, and Isopycnals are surfaces of equal water density.

Geology

Various geological data are rendered as contour maps in structural geology, sedimentology, stratigraphy and economic geology. Contour maps are used to show the below ground surface of geologic strata, fault surfaces (especially low angle thrust faults) and unconformities. Isopach maps use isopachs (lines of equal thickness) to illustrate variations in thickness of geologic units.

Environmental Science

In discussing pollution, density maps can be very useful in indicating sources and areas of greatest contamination. Contour maps are especially useful for diffuse forms or scales of pollution. Acid precipitation is indicated on maps with isoplats. Some of the most widespread applications of environmental science contour maps involve mapping of environmental noise (where lines of equal sound pressure level are denoted isobels), air pollution, soil contamination, thermal pollution and groundwater contamination. By contour planting and contour ploughing, the rate of water runoff and thus soil erosion can be substantially reduced; this is especially important in riparian zones.

Ecology

An isoflor is an isopleth contour connecting areas of comparable biological diversity. Usually, the variable is the number of species of a given genus or family that occurs in a region. Isoflor maps are thus used to show distribution patterns and trends such as centres of diversity.

Social Sciences

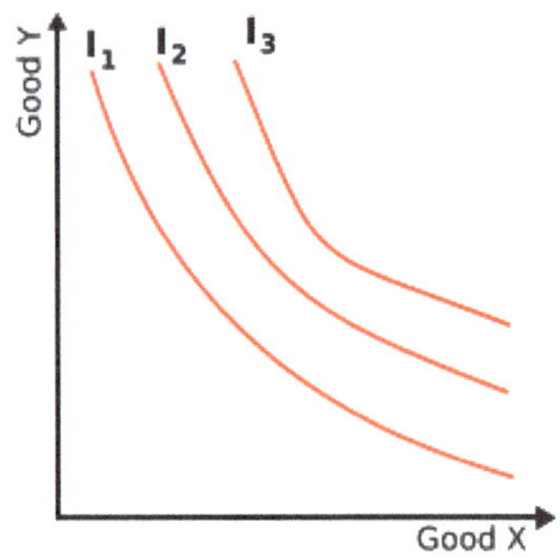

From economics, an indifference map with three indifference curves shown. All points on a particular indifference curve have the same value of the utility function, whose values implicitly come out of the page in the unshown third dimension.

In economics, contour lines can be used to describe features which vary quantitatively over space. An isochrone shows lines of equivalent drive time or travel time to a given location and is used in the generation of isochrone maps. An isotim shows equivalent transport costs from the source of a raw material, and an isodapane shows equivalent cost of travel time.

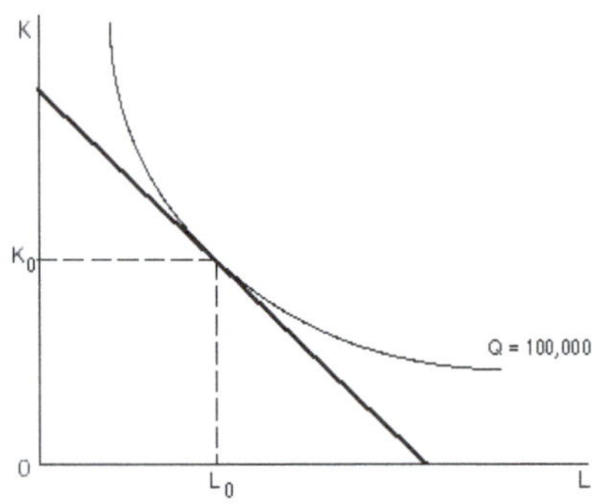

A single production isoquant (convex) and a single isocost curve (linear). Labor usage is plotted horizontally and physical capital usage is plotted vertically.

Contour lines are also used to display non-geographic information in economics. **In**difference curves (as shown at left) are used to show bundles of goods to which a person would assign equal utility. An isoquant (in the image at right) is a curve of equal production quantity for alternative combinations of input usages, and an isocost curve (also in the image at right) shows alternative combinations of input usages having equal production costs.

In political science an analogous method is used in understanding coalitions (for example the diagram in Laver and Shepsle's work).

In population dynamics, an isocline shows the set of population sizes at which the rate of change, or partial derivative, for one population in a pair of interacting populations is zero.

Statistics

In statistics, isodensity lines or isodensanes are lines that joint points with the same probability density. Isodensanes are used to display bivariate distributions.

Thermodynamics, engineering, and other sciences

Various types of graphs in thermodynamics, engineering, and other sciences use isobars (constant pressure), isotherms (constant temperature), isochors (constant specific volume), or other types of isolines, even though these graphs are usually not related to maps. Such isolines are useful for representing more than two dimensions (or quantities) on two-dimensional graphs. Common examples in thermodynamics are some types of phase diagrams.

Isoclines are used to solve ordinary differential equations.

In interpreting radar images, an isodop is a line of equal Doppler velocity, and an isoecho is a line of equal radar reflectivity.

Other Phenomena

- isochasm: aurora equal occurrence
- isochor: volume
- isodose: Absorbed dose of radiation
- isophene: biological events occurring with coincidence such as plants flowering
- isophote: illuminance

History

The idea of lines that join points of equal value was rediscovered several times. In 1701, Edmond Halley used such lines (isogons) on a chart of magnetic variation. The Dutch engineer Nicholas Cruquius drew the bed of the river Merwede with lines of equal depth (isobaths) at intervals of 1 fathom in 1727, and Philippe Buache used them at 10-fathom intervals on a chart of the English Channel that was prepared in 1737 and published in 1752. Such lines were used to describe a land surface (contour lines) in a map of the Duchy of Modena and Reggio by Domenico Vandelli in 1746, and they were studied theoretically by Ducarla in 1771, and Charles Hutton used them in the Schiehallion experiment. In 1791, a map of France by J. L. Dupain-Triel used contour lines at 20-metre intervals, hachures, spot-heights and a vertical section. In 1801, the chief of the Corps of Engineers, Haxo, used contour lines at the larger scale of 1:500 on a plan of his projects for Rocca d'Aufo.

By around 1843, when the Ordnance Survey started to regularly record contour lines in Great Britain and Ireland, they were already in general use in European countries. Isobaths were not routinely used on nautical charts until those of Russia from 1834, and those of Britain from 1838.

When maps with contour lines became common, the idea spread to other applications. Perhaps the latest to develop are air quality and noise pollution contour maps, which first appeared in the US, in approximately 1970, largely as a result of national legislation requiring spatial delineation of these parameters. In 2007, Pictometry International was the first to allow users to dynamically generate elevation contour lines to be laid over oblique images.

Graphical Design

To maximize readability of contour maps, there are several design choices available to the map creator, principally line weight, line color, line type and method of numerical marking.

Line weight is simply the darkness or thickness of the line used. This choice is made based upon the least intrusive form of contours that enable the reader to decipher the background information in the map itself. If there is little or no content on the base map, the contour lines may be drawn with relatively heavy thickness. Also, for many forms of contours such as topographic maps, it is common to vary the line weight and/ or color, so that a different line characteristic occurs for certain numerical values. For example, in the topographic map above, the even hundred foot elevations are shown in a different weight from the twenty foot intervals.

Line color is the choice of any number of pigments that suit the display. Sometimes a sheen or gloss is used as well as color to set the contour lines apart from the base map. Line colour can be varied to show other information.

Line type refers to whether the basic contour line is solid, dashed, dotted or broken in some other pattern to create the desired effect. Dotted or dashed lines are often used when the underlying base map conveys very important (or difficult to read) information. Broken line types are used when the location of the contour line is inferred.

Numerical marking is the manner of denoting the arithmetical values of contour lines. This can be done by placing numbers along some of the contour lines, typically using interpolation for intervening lines. Alternatively a map key can be produced associating the contours with their values.

If the contour lines are not numerically labeled and adjacent lines have the same style (with the same weight, color and type), then the direction of the gradient cannot be determined from the contour lines alone. However, if the contour lines cycle through three or more styles, then the direction of the gradient can be determined from the lines. The orientation of the numerical text labels is often used to indicate the direction of the slope.

Plan view versus profile view

Most commonly contour lines are drawn in plan view, or as an observer in space would view the Earth's surface: ordinary map form. However, some parameters can often be displayed in profile view showing a vertical profile of the parameter mapped. Some of the most common parameters mapped in profile are air pollutant concentrations and sound levels. In each of those cases it may be important to analyze (air pollutant concentrations or sound levels) at varying heights so as to determine the air quality or noise health effects on people at different elevations, for example, living on different floor levels of an urban apartment. In actuality, both plan and profile view contour maps are used in air pollution and noise pollution studies.

Labeling contour maps

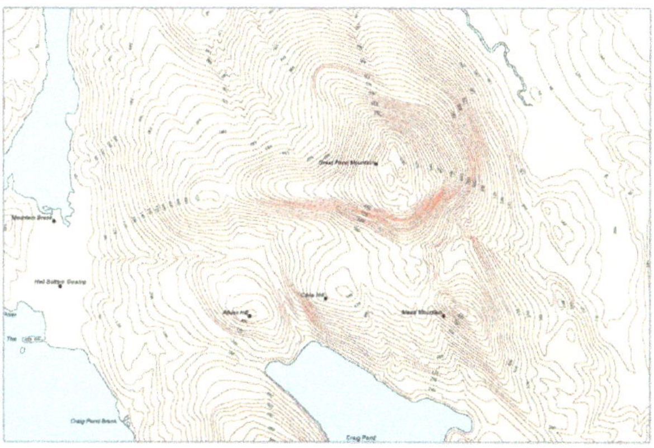

Contour map labeled aesthetically in an "elevation up" manner.

Labels are a critical component of elevation maps. A properly labeled contour map helps the reader to quickly interpret the shape of the terrain. If numbers are placed close to each other, it means that the terrain is steep. Labels should be placed along a slightly curved line "pointing" to the summit or nadir, from several directions if possible, making the visual identification of the summit or nadir easy. Contour labels can be oriented so a reader is facing uphill when reading the label.

Manual labeling of contour maps is a time-consuming process, however, there are a few software systems that can do the job automatically and in accordance with cartographic conventions, called automatic label placement.

References

- Kurt A. Raaflaub; Richard J. A. Talbert (2009). Geography and Ethnography: Perceptions of the World in Pre-Modern Societies. John Wiley & Sons. p. 147. ISBN 1-4051-9146-5.
- Miyajima, Kazuhiko (1997). "Projection methods in Chinese, Korean and Japanese star maps". In Johannes Andersen. Highlights of Astronomy. 11B. Norwell: Kluwer Academic Publishers. p. 714. ISBN 978-0-7923-5556-4.
- Needham, Joseph (1971). Part 3: Civil Engineering and Nautics. Science and Civilization in China. 4. Cambridge University Press. p. 569. ISBN 978-0-521-07060-7.
- Sircar, D. C. C. (1990). Studies in the Geography of Ancient and Medieval India. Motilal Banarsidass Publishers. p. 330. ISBN 81-208-0690-5.
- Devlin, Keith (2002). The Millennium Problems. New York, New York: Basic Books. pp. 162–163. ISBN 978-0-465-01730-0.
- Robinson, A.H. (1982). Early Thematic Mapping: In the History of Cartography. Chicago: The University of Chicago Press. ISBN 0-226-72285-6.
- de Crespigny, Rafe. (2007). A Biographical Dictionary of Later Han to the Three Kingdoms (23–220 AD). Leiden: Koninklijke Brill. ISBN 90-04-15605-4.
- Ebrey, Patricia Buckley, Anne Walthall, James B. Palais (2006). East Asia: A Cultural, Social, and Political History. Boston: Houghton Mifflin Company. ISBN 0-618-13384-4.
- Samuel Noah Kramer: Der Stadtplan von Nippur, der älteste Stadtplan der Welt. In Helmut Uhlig: Die Sumerer. Lübbe, Bergisch-Gladbach 1992. ISBN 3-404-64117-5.
- Introduction by Prof.H. J. Dyos to Collins' Illustrated Atlas of London, 1973 edition, Leicester University Press, ISBN 0-7185-5012-9:
- Snyder, John P. (1993). Flattening the earth: two thousand years of map projections. University of Chicago Press. ISBN 0-226-76746-9.
- Choosing a World Map. Falls Church, Virginia: American Congress on Surveying and Mapping. 1988. p. 1. ISBN 0-9613459-2-6.
- Slocum, Terry A.; Robert B. McMaster; Fritz C. Kessler; Hugh H. Howard (2005). Thematic Cartography and Geographic Visualization (2nd ed.). Upper Saddle River, NJ: Pearson Prentice Hall. p. 166. ISBN 0-13-035123-7.
- T. Slocum, R. McMaster, F. Kessler, and H. Howard, Thematic Cartography and Geographic Visualization, 2nd edition, Pearson, 2005, ISBN 0-13-035123-7, p. 272.
- Guanqun, Wang. "China issues new rules on Internet map publishing". news.xinhuanet.com. Xinhua News Agency. Retrieved 27 July 2016

- Government of Canada (2016-04-08). "National Topographic System Maps". Earth Sciences – Geography. Natural Resources Canada. Retrieved 2016-05-16.
- "Maclure's geological map of the United States". US Library of Congress' Map Collection. Library of Congress. Retrieved 30 October 2015.
- "Forecasting weather system movement with pressure tendency". Chapter 13 - Weather Forecasting. Lyndon State College Atmospheric Sciences. Retrieved 12 April 2014.
- Campbell-Dollaghan, Kelsey (July 22, 2013). "7 Brilliant Reinventions of Buckminster Fuller's Dymaxion Map". Gizmodo. Retrieved 21 January 2014.
- Strange Maps (Mar 30, 2009). "370 – Palestine's Island Paradise, Now With a Word from its Creator". Blog. StrangeMaps. Retrieved 2013-05-16.

6

Allied Fields of Topography

Topography is an interdisciplinary subject. To develop a better understanding of topography, its allied fields also need to be contemplated. Earth science studies the planet Earth which includes the interior as well as the exterior of the planet. Planetary science is the study of planets and the process that helps in their formation. This chapter will provide a glimpse of the allied fields of topography.

Earth Science

Earth science or geoscience is an all-embracing term for the fields of science related to the planet Earth. It can be considered to be a branch of planetary science, but with a much older history. There are both reductionist and holistic approaches to Earth sciences. The Earth sciences can include the study of geology, the lithosphere, and the large-scale structure of the Earth's interior, as well as the atmosphere, hydrosphere, and biosphere. Typically, Earth scientists use tools from geography, physics, chemistry, biology, chronology, and mathematics to build a quantitative understanding of how the Earth system works and evolves.

Fields of Study

The following fields of science are generally categorized within the Earth sciences:

- Physical geography, covers aspects of geomorphology, soil study, hydrology, meteorology, climatology, and biogeography.
- Geology describes the rocky parts of the Earth's crust (or lithosphere) and its historic development. Major subdisciplines are mineralogy and petrology, geochemistry, geomorphology, paleontology, stratigraphy, structural geology, engineering geology, and sedimentology.
- Geophysics and geodesy investigate the shape of the Earth, its reaction to forces and its magnetic and gravity fields. Geophysicists explore the Earth's core and mantle as well as the tectonic and seismic activity of the lithosphere. Geophysics is commonly used to supplement the work of geologists in developing a comprehensive understanding of crustal geology, particularly in mineral and petroleum exploration.

- Soil science covers the outermost layer of the Earth's crust that is subject to soil formation processes (or pedosphere). Major subdisciplines include edaphology and pedology.
- Ecology covers the interactions between the biota, with their natural environment. This field of study differentiates the study of the Earth, from the study of other planets in the Solar System; the Earth being the only planet teeming with life.
- Hydrology (includes oceanography and limnology) describe the marine and freshwater domains of the watery parts of the Earth (or hydrosphere). Major subdisciplines include hydrogeology and physical, chemical, and biological oceanography.
- Glaciology covers the icy parts of the Earth (or cryosphere).
- Atmospheric sciences cover the gaseous parts of the Earth (or atmosphere) between the surface and the exosphere (about 1000 km). Major subdisciplines include meteorology, climatology, atmospheric chemistry, and atmospheric physics.

Earth's Interior

A volcanic eruption is the release of stored energy from below the surface of Earth.

Plate tectonics, mountain ranges, volcanoes, and earthquakes are geological phenomena that can be explained in terms of energy transformations in Earth's crust.

Beneath Earth's crust lies the mantle which is heated by the radioactive decay of heavy elements. The mantle is not quite solid and consists of magma which is in a state of semi-perpetual convection. This convection process causes the lithospheric plates to move, albeit slowly. The resulting process is known as plate tectonics.

Plate tectonics might be thought of as the process by which the Earth is resurfaced. Through a process called seafloor spreading, new crust is created by the flow of magma from underneath the lithosphere to the surface, through fissures, where it cools and solidifies. Through a process called subduction, oceanic crust is pushed underground — beneath the rest of the lithosphere—where it comes into contact with magma and melts—rejoining the mantle from which it originally came.

Areas of the crust where new crust is created are called *divergent boundaries*, those where it is brought back into the Earth are *convergent boundaries* and those where plates slide past each other, but no new lithospheric material is created or destroyed, are referred to as transform (or conservative) boundaries Earthquakes result from the movement of the lithospheric plates, and they often occur near convergent boundaries where parts of the crust are forced into the Earth as part of subduction.

Volcanoes result primarily from the melting of subducted crust material. Crust material that is forced into the asthenosphere melts, and some portion of the melted material becomes light enough to rise to the surface—giving birth to volcanoes.

Earth's Electromagnetic Field

An electromagnet is a magnet that is created by a current that flows around a soft iron core. Earth has a solid iron inner core surrounded by semi-liquid materials of the outer core that move in continuous currents around the inner core; therefore, the Earth is an electromagnet. This is referred to as the dynamo theory of Earth's magnetism.

Earth'S Atmosphere

The magnetosphere shields the surface of Earth from the charged particles of the solar wind. (*image not to scale.*)

The troposphere, stratosphere, mesosphere, thermosphere, and exosphere are the five layers which make up Earth's atmosphere. In all, the atmosphere is made up of about 78.0% nitrogen, 20.9% oxygen, and 0.92% argon. 75% of the gases in the atmosphere are located within the troposphere, the bottom-most layer. The remaining one percent of the atmosphere (all but the nitrogen, oxygen, and argon) contains small amounts of other gases including CO_2 and water vapors. Water vapors and CO_2 allow the Earth's atmosphere to catch and hold the Sun's energy through a phenomenon called the greenhouse effect. This allows Earth's surface to be warm enough to have liquid water and support life.

The magnetic field created by the internal motions of the core produces the magnetosphere which protects the Earth's atmosphere from the solar wind. As the Earth is 4.5 billion years old, it would have lost its atmosphere by now if there were no protective magnetosphere.

In addition to storing heat, the atmosphere also protects living organisms by shielding the Earth's surface from cosmic rays. Note that the level of protection is high enough to prevent cosmic rays from destroying all life on Earth, yet low enough to aid the mutations that have an important role in pushing forward diversity in the biosphere.

Methodology

Methodologies vary depending on the nature of the subjects being studied. Studies typically fall into one of three categories: observational, experimental, or theoretical. Earth scientists often conduct sophisticated computer analysis or go to many of the world's most exotic locations to study Earth phenomena (e.g. Antarctica or hot spot island chains).

A foundational idea within the study Earth science is the notion of uniformitarianism. Uniformitarianism dictates that "ancient geologic features are interpreted by understanding active processes that are readily observed." In other words, any geologic processes at work in the present have operated in the same ways throughout geologic time. This enables those who study Earth's history to apply knowledge of how Earth processes operate in the present to gain insight into how the planet has evolved and changed throughout deep history.

Earth's spheres

Earth science generally recognizes four spheres, the lithosphere, the hydrosphere, the atmosphere, and the biosphere; these correspond to rocks, water, air and life. Also included by some are the cryosphere (corresponding to ice) as a distinct portion of the hydrosphere and the pedosphere (corresponding to soil) as an active and intermixed sphere.

Partial list of the major earth science topics

Atmosphere

- Atmospheric chemistry
- Geography
    - Climatology
    - Meteorology
- Hydrometeorology
- Paleoclimatology

Biosphere

- Biogeochemistry
- Biogeography

- Ecology
    - Landscape ecology
- Geoarchaeology
- Geomicrobiology
- Paleontology
    - Palynology
    - Micropaleontology

Hydrosphere

- Hydrology
    - Hydrogeology
- Limnology (freshwater science)
- Oceanography (marine science)
    - Chemical oceanography
    - Physical oceanography
    - Biological oceanography (marine biology)
    - Geological oceanography (marine geology)
        - Paleoceanography

Lithosphere (Geosphere)

- Geology
    - Economic geology
    - Engineering geology
    - Environmental geology
    - Historical geology
        - Quaternary geology
    - Planetary geology and planetary geography
    - Sedimentology
    - Stratigraphy

    - Structural geology
- Geography
    - Physical geography
- Geochemistry
- Geomorphology
- Geophysics
    - Geochronology
    - Geodynamics
    - Geomagnetism
    - Gravimetry (also part of Geodesy)
    - Seismology
- Glaciology
- Hydrogeology
- Mineralogy
    - Crystallography
    - Gemology
- Petrology
- Speleology
- Volcanology

Pedosphere

- Geography
    - Soil science
        - Edaphology
        - Pedology

Systems

- Earth system science
- Environmental science

- Geography
    - Human geography
    - Physical geography
- Gaia hypothesis
- Systems ecology
- Systems geology

Others

- Geography
    - Cartography
    - Geoinformatics (GIS)
    - Geostatistics
    - Geodesy and Surveying
    - NASA Earth Science Enterprise
    - Remote Sensing

Planetary Science

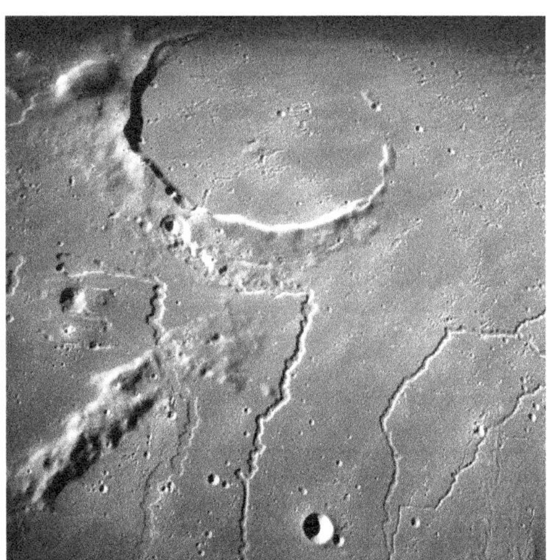

Photograph from Apollo 15 orbital unit of the rilles in the vicinity of the crater Aristarchus on the Moon. The arrangement of the two valleys is very similar, although one third the size, to Great Hungarian Plain rivers Danube and Tisza.

There are interrelated observational and theoretical branches of planetary science. Observational research can involve a combination of space exploration, predominantly with robotic spacecraft missions using remote sensing, and comparative, experimental work in Earth-based laboratories. The theoretical component involves considerable computer simulation and mathematical modelling.

Planetary scientists are generally located in the astronomy and physics or Earth sciences departments of universities or research centres, though there are several purely planetary science institutes worldwide. There are several major conferences each year, and a wide range of peer-reviewed journals.

History

The history of planetary science may be said to have begun with the Ancient Greek philosopher Democritus, who is reported by Hippolytus as saying

The ordered worlds are boundless and differ in size, and that in some there is neither sun nor moon, but that in others, both are greater than with us, and yet with others more in number. And that the intervals between the ordered worlds are unequal, here more and there less, and that some increase, others flourish and others decay, and here they come into being and there they are eclipsed. But that they are destroyed by colliding with one another. And that some ordered worlds are bare of animals and plants and all water.

In more modern times, planetary science began in astronomy, from studies of the unresolved planets. In this sense, the original planetary astronomer would be Galileo, who discovered the four largest moons of Jupiter, the mountains on the Moon, and first observed the rings of Saturn, all objects of intense later study. Galileo's study of the lunar mountains in 1609 also began the study of extraterrestrial landscapes: his observation "that the Moon certainly does not possess a smooth and polished surface" suggested that it and other worlds might appear "just like the face of the Earth itself".

Advances in telescope construction and instrumental resolution gradually allowed increased identification of the atmospheric and surface details of the planets. The Moon was initially the most heavily studied, as it always exhibited details on its surface, due to its proximity to the Earth, and the technological improvements gradually produced more detailed lunar geological knowledge. In this scientific process, the main instruments were astronomical optical telescopes (and later radio telescopes) and finally robotic exploratory spacecraft.

The Solar System has now been relatively well-studied, and a good overall understanding of the formation and evolution of this planetary system exists. However, there are large numbers of unsolved questions, and the rate of new discoveries is very high, partly due to the large number of interplanetary spacecraft currently exploring the Solar System.

Disciplines

Planetary Astronomy

This is both an observational and a theoretical science. Observational researchers are predominantly concerned with the study of the small bodies of the Solar System: those that are observed by telescopes, both optical and radio, so that characteristics of these bodies such as shape, spin, surface materials and weathering are determined, and the history of their formation and evolution can be understood.

Theoretical planetary astronomy is concerned with dynamics: the application of the principles of celestial mechanics to the Solar System and extrasolar planetary systems.

Planetary Geology

The best known research topics of planetary geology deal with the planetary bodies in the near vicinity of the Earth: the Moon, and the two neighbouring planets: Venus and Mars. Of these, the Moon was studied first, using methods developed earlier on the Earth.

Geomorphology

Geomorphology studies the features on planetary surfaces and reconstructs the history of their formation, inferring the physical processes that acted on the surface. Planetary geomorphology includes study of several classes of surface feature:

- Impact features (multi-ringed basins, craters)
- Volcanic and tectonic features (lava flows, fissures, rilles)
- Space weathering - erosional effects generated by the harsh environment of space (continuous micrometeorite bombardment, high-energy particle rain, impact gardening). For example, the thin dust cover on the surface of the lunar regolith is a result of micrometeorite bombardment.
- Hydrological features: the liquid involved can range from water to hydrocarbon and ammonia, depending on the location within the Solar System.

The history of a planetary surface can be deciphered by mapping features from top to bottom according to their deposition sequence, as first determined on terrestrial strata by Nicolas Steno. For example, stratigraphic mapping prepared the Apollo astronauts for the field geology they would encounter on their lunar missions. Overlapping sequences were identified on images taken by the Lunar Orbiter program, and these were used to prepare a lunar stratigraphic column and geological map of the Moon.

Cosmochemistry, Geochemistry and Petrology

One of the main problems when generating hypotheses on the formation and evolution of objects in the Solar System is the lack of samples that can be analysed in the laboratory, where a large suite of tools are available and the full body of knowledge derived from terrestrial geology can be brought to bear. Fortunately, direct samples from the Moon, asteroids and Mars are present on Earth, removed from their parent bodies and delivered as meteorites. Some of these have suffered contamination from the oxidising effect of Earth's atmosphere and the infiltration of the biosphere, but those meteorites collected in the last few decades from Antarctica are almost entirely pristine.

The different types of meteorite that originate from the asteroid belt cover almost all parts of the structure of differentiated bodies: meteorites even exist that come from the core-mantle boundary (pallasites). The combination of geochemistry and observational astronomy has also made it possible to trace the HED meteorites back to a specific asteroid in the main belt, 4 Vesta.

The comparatively few known Martian meteorites have provided insight into the geochemical composition of the Martian crust, although the unavoidable lack of information about their points of origin on the diverse Martian surface has meant that they do not provide more detailed constraints on theories of the evolution of the Martian lithosphere. As of July 24, 2013 65 samples of Martian meteorites have been discovered on Earth. Many were found in either Antarctica or the Sahara Desert.

During the Apollo era, in the Apollo program, 384 kilograms of lunar samples were collected and transported to the Earth, and 3 Soviet Luna robots also delivered regolith samples from the Moon. These samples provide the most comprehensive record of the composition of any Solar System body beside the Earth. The numbers of lunar meteorites are growing quickly in the last few years – as of April 2008 there are 54 meteorites that have been officially classified as lunar. Eleven of these are from the US Antarctic meteorite collection, 6 are from the Japanese Antarctic meteorite collection, and the other 37 are from hot desert localities in Africa, Australia, and the Middle East. The total mass of recognized lunar meteorites is close to 50 kg.

Geophysics

Space probes made it possible to collect data in not only the visible light region, but in other areas of the electromagnetic spectrum. The planets can be characterized by their force fields: gravity and their magnetic fields, which are studied through geophysics and space physics.

Measuring the changes in acceleration experienced by spacecraft as they orbit has allowed fine details of the gravity fields of the planets to be mapped. For example, in the 1970s, the gravity field disturbances above lunar maria were measured through lunar orbiters, which led to the discovery of concentrations of mass, mascons, beneath the Imbrium, Serenitatis, Crisium, Nectaris and Humorum basins.

If a planet's magnetic field is sufficiently strong, its interaction with the solar wind forms a magnetosphere around a planet. Early space probes discovered the gross dimensions of the terrestrial magnetic field, which extends about 10 Earth radii towards the Sun. The solar wind, a stream of charged particles, streams out and around the terrestrial magnetic field, and continues behind the magnetic tail, hundreds of Earth radii downstream. Inside the magnetosphere, there are relatively dense regions of solar wind particles, the Van Allen radiation belts.

Geophysics includes seismology and tectonophysics, geophysical fluid dynamics, mineral physics, geodynamics, mathematical geophysics, and geophysical surveying.

Geodesy, also called geodetics, deals with the measurement and representation of the planets of the Solar System, their gravitational fields and geodynamic phenomena (polar motion in three-dimensional, time-varying space. The science of geodesy has elements of both astrophysics and planetary sciences. The shape of the Earth is to a large extent the result of its rotation, which causes its equatorial bulge, and the competition of geologic processes such as the collision of plates and of vulcanism, resisted by the Earth's gravity field.

These principles can be applied to the solid surface of Earth (orogeny; Few mountains are higher than 10 km (6 mi), few deep sea trenches deeper than that because quite simply, a mountain as tall as, for example, 15 km (9 mi), would develop so much pressure at its base, due to gravity, that the rock there would become plastic, and the mountain would slump back to a height of roughly 10 km (6 mi) in a geologically insignificant time. Some or all of these geologic principles can be applied to other planets besides Earth. For instance on Mars, whose surface gravity is much less, the largest volcano, Olympus Mons, is 27 km (17 mi) high at its peak, a height that could not be maintained on Earth. The Earth geoid is essentially the figure of the Earth abstracted from its topographic features. Therefore, the Mars geoid is essentially the figure of Mars abstracted from its topographic features. Surveying and mapping are two important fields of application of geodesy.

Atmospheric Science

The atmosphere is an important transitional zone between the solid planetary surface and the higher rarefied ionizing and radiation belts. Not all planets have atmospheres: their existence depends on the mass of the planet, and the planet's distance from the Sun — too distant and frozen atmospheres occur. Besides the four gas giant planets, almost all of the terrestrial planets (Earth, Venus, and Mars) have significant atmospheres. Two moons have significant atmospheres: Saturn's moon Titan and Neptune's moon Triton. A tenuous atmosphere exists around Mercury.

Cloud bands clearly visible on Jupiter.

The effects of the rotation rate of a planet about its axis can be seen in atmospheric streams and currents. Seen from space, these features show as bands and eddies in the cloud system, and are particularly visible on Jupiter and Saturn.

Comparative Planetary Science

Planetary science frequently makes use of the method of comparison to give greater understanding of the object of study. This can involve comparing the dense atmospheres of Earth and Saturn's moon Titan, the evolution of outer Solar System objects at different distances from the Sun, or the geomorphology of the surfaces of the terrestrial planets, to give only a few examples.

The main comparison that can be made is to features on the Earth, as it is much more accessible and allows a much greater range of measurements to be made. Earth analogue studies are particularly common in planetary geology, geomorphology, and also in atmospheric science.

Professional Activity

Professional Bodies

- Division for Planetary Sciences (DPS) of the American Astronomical Society
- American Geophysical Union
- Meteoritical Society
- Europlanet

Major Conferences

- Lunar and Planetary Science Conference (LPSC), organized by the Lunar and Planetary Institute in Houston. Held annually since 1970, occurs in March.
- Division for Planetary Sciences (DPS) meeting held annually since 1970 at a different location each year, predominantly within the mainland US. Occurs around October.
- American Geophysical Union (AGU) annual Fall meeting in December in San Francisco.
- American Geophysical Union (AGU) Joint Assembly (co-sponsored with other societies) in April–May, in various locations around the world.
- Meteoritical Society annual meeting, held during the Northern Hemisphere summer, generally alternating between North America and Europe.
- European Planetary Science Congress (EPSC), held annually around September at a location within Europe.

Smaller workshops and conferences on particular fields occur worldwide throughout the year.

Major Institutions

This non-exhaustive list includes those institutions and universities with major groups of people working in planetary science. Alphabetical order is used.

National Space Agencies

- Ames (NASA)
- Canadian Space Agency (CSA). Annual budget CAD $488.7 million (2013–2014).
- China National Space Administration (CNSA) (People's Republic of China). Budget $0.5-1.3 Billion (est.).
- Centre national d'études spatiales French National Centre of Space Research,Budget €1.920 Billion (2012).
- Deutsches Zentrum für Luft- und Raumfahrt e.V., (German: abbreviated DLR), the German Aerospace Center.Budget $2 Billion (2010).
- European Space Agency (ESA). Budget $5.51 Billion (2013).
- Russian Federal Space Agency Budget $5.61 Billion (2013).

- GSFC (NASA),
- Indian Space Research Organisation (ISRO),
- Israel Space Agency (ISA),
- Italian Space Agency Budget ~$1 Billion (2010).
- Japan Aerospace Exploration Agency (JAXA). Budget $2.15 Billion (2012).
- JPL (NASA),
- NASA: Considerable number of research groups, including the JPL, GSFC, Ames. Budget $18.724 Billion (2011).
- National Space Organization (Republic of China in Taiwan).
- UK Space Agency (UKSA).

Other Institutions

- The Australian National University's Planetary Science Institute
- Brown University Planetary Geosciences Group
- Caltech's Division of Geological and Planetary Sciences
- Cornell University's Space and Planetary Science
- Florida Institute of Technology's Department of Physics and Space Sciences
- Johns Hopkins University Applied Physics Laboratory
- Lunar and Planetary Institute
- Max Planck Institute for Solar System Research's Department Planets and Comets
- MIT Dept. of Earth, Atmospheric and Planetary Sciences
- Open University Planetary and Space Sciences Research Institute
- Planetary Science Institute
- Stony Brook University's Geosciences Department and soon to open Center for Planetary Exploration
- UCL/Birkbeck's Centre for Planetary Sciences
- UCLA Dept. of Earth and Space Sciences
- University of Arizona's Lunar and Planetary Lab
- University of California Santa Cruz's Department of Earth & Planetary Sciences

- University of Hawaii's Hawaii Institute of Geophysics and Planetology
- University of Copenhagen's Center for Planetary Research
- University of Central Florida Planetary Sciences Group
- University of British Columbia Institute for Planetary Science
- University of Western Ontario's Centre for Planetary Science and Exploration
- University of Tennessee Department of Earth and Planetary Sciences
- University of Colorado's Department of Astrophysical and Planetary Sciences
- INAF– Istituto di Astrofisica e Planetologia Spaziali

Basic Concepts

- Asteroid
- Celestial mechanics
- Comet
- Dwarf planet
- Extrasolar planet
- Gas giant
- Icy moon
- Kuiper belt
- Magnetosphere
- Minor planet
- Planet
- Planetary differentiation
- Planetary system
- Definition of a planet
- Space weather
- Terrestrial planet

References

- Demorest, Paul (2001-05-21). "Dynamo Theory and Earth's Magnetic Field." (PDF). Archived from the original (PDF) on February 21, 2007. Retrieved 2007-11-17.
- Earth's Spheres. ©1997-2000. Wheeling Jesuit University/NASA Classroom of the Future. Retrieved November 11, 2007.
- Hippolytus (Antipope); Origen (1921). Philosophumena (Digitized 9 May 2006). 1. Translation by Francis Legge, F.S.A. Original from Harvard University.: Society for promoting Christian knowledge. Retrieved 22 May 2009.

Permissions

All chapters in this book are published with permission under the Creative Commons Attribution Share Alike License or equivalent. Every chapter published in this book has been scrutinized by our experts. Their significance has been extensively debated. The topics covered herein carry significant information for a comprehensive understanding. They may even be implemented as practical applications or may be referred to as a beginning point for further studies.

We would like to thank the editorial team for lending their expertise to make the book truly unique. They have played a crucial role in the development of this book. Without their invaluable contributions this book wouldn't have been possible. They have made vital efforts to compile up to date information on the varied aspects of this subject to make this book a valuable addition to the collection of many professionals and students.

This book was conceptualized with the vision of imparting up-to-date and integrated information in this field. To ensure the same, a matchless editorial board was set up. Every individual on the board went through rigorous rounds of assessment to prove their worth. After which they invested a large part of their time researching and compiling the most relevant data for our readers.

The editorial board has been involved in producing this book since its inception. They have spent rigorous hours researching and exploring the diverse topics which have resulted in the successful publishing of this book. They have passed on their knowledge of decades through this book. To expedite this challenging task, the publisher supported the team at every step. A small team of assistant editors was also appointed to further simplify the editing procedure and attain best results for the readers.

Apart from the editorial board, the designing team has also invested a significant amount of their time in understanding the subject and creating the most relevant covers. They scrutinized every image to scout for the most suitable representation of the subject and create an appropriate cover for the book.

The publishing team has been an ardent support to the editorial, designing and production team. Their endless efforts to recruit the best for this project, has resulted in the accomplishment of this book. They are a veteran in the field of academics and their pool of knowledge is as vast as their experience in printing. Their expertise and guidance has proved useful at every step. Their uncompromising quality standards have made this book an exceptional effort. Their encouragement from time to time has been an inspiration for everyone.

The publisher and the editorial board hope that this book will prove to be a valuable piece of knowledge for students, practitioners and scholars across the globe.

Index

V

W